ENERGY POLICIES, POLITICS AND PRICES SERIES

POWER PLANT CHARACTERISTICS AND COSTS

Energy Policies, Politics and Prices Series

Nuclear Power's Role in Generating Electricity
Perry G. Furham
2009. ISBN: 978-1-60741-226-7

OPEC, Oil Prices and LNG
Edward R. Pitt and Christopher N. Leung (Editors)
2009. ISBN: 978-1-60692-897-4

OPEC, Oil Prices and LNG
Edward R. Pitt and Christopher N. Leung (Editors)
2009. ISBN: 978-1-60876-614-7 (Online Book)

Dynamic Noncooperative Game Models for Deregulated Electricity Markets
Jose B. Cruz, Jr. and Xiaohuan Tan (Editors)
2009. ISBN: 978-1-60741-078-2

Energy Prices: Supply, Demand or Speculation?
John T. Perry (Editor)
2009. ISBN: 978-1-60741-374-5

Worldwide Biomass Potential: Technology Characterizations
R. L. Bain
2010. ISBN: 978-1-60741-267-0

Power Plant Characteristics and Costs
Stan Kaplan
2010. ISBN: 978-1-60741-264-9

ENERGY POLICIES, POLITICS AND PRICES SERIES

POWER PLANT CHARACTERISTICS AND COSTS

Stan Kaplan

Nova Science Publishers, Inc.
New York

For permission to use material from this book please contact us:
Telephone 631-231-7269; Fax 631-231-8175
Web Site: http://www.novapublishers.com

LIBRARY OF CONGRESS CATALOGING-IN-PUBLICATION DATA

Kaplan, Stan, 1947-
Power plant characteristics and costs / Stan Kaplan.
p. cm. Includes index.
ISBN 978-1-60741-264-9 (hardcover)
1. Electric power-plants--United States. 2. Electric power-plants--United States--Cost of operation. 3. Electric power-plants--United States--Design and construction. 4. Electric power production--United States--Forecasting. I. Title.
TK1193.U5K287 2009
621.31'210973--dc22 2009028850

Published by Nova Science Publishers, Inc. ✢ New York

CONTENTS

PREFACE

This is an edited, excerpted and augmented edition of a United States Congressional Research Service publication, Report Order Code RL34746, dated November 13, 2008, by Stan Kaplan, Specialist in Energy and Environmental Policy Resources, Science, and Industry Division

This book analyzes the factors that determine the cost of electricity from new power plants. These factors, including construction costs, fuel expense, environmental regulations, and financing costs can all be affected by government energy, environmental, and economic policies. Government decisions to influence or not influence these factors can largely determine the kind of power plants that are built in the future. This book provides projections of the possible cost of power from new fossil, nuclear, and renewable plants built in 2015, illustrating how different assumptions, such as the availability of federal incentives, change the cost rankings of technologies.

None of the projections are intended to be a “most likely” case. Future uncertainties preclude firm forecasts. The rankings of the technologies by cost are therefore also an approximation and should not be viewed as definitive estimates of the relative cost-competitiveness of each option. The value of this book is not as a source of point estimates of future power costs, but as a source of insight into the factors that can determine future outcomes, including factors that can be influenced by Congress.

Chapter 1

INTRODUCTION AND ORGANIZATION

The United States may have to build many new power plants to meet growing demand for electric power. For example, the Energy Information Administration (EIA) estimates that the nation will have to construct 226,000 megawatts of new electric power generating capacity by 2030.[1] This is the equivalent of about 450 large power plants. Whatever the number of plants actually built, different combinations of fossil, nuclear, or renewable plants could be built to meet the demand for new generating capacity. Congress can largely determine which kinds of plants are actually built through energy, environmental, and economic policies that influence power plant costs.

This report analyzes the factors that determine the cost of electricity from new power plants. These factors — including construction costs, fuel expense, environmental regulations, and financing costs — can all be affected by government energy and economic policies. Government decisions to influence, or not influence, these factors can largely determine the kind of power plants that are built in the future. For example, government policies aimed at reducing the cost of constructing power plants could especially benefit nuclear plants, which are costly to build. Policies that reduce the cost of fossil fuels could benefit natural gas plants, which are inexpensive to build but rely on an expensive fuel.

The report provides projections of the possible cost of power for new fossil, nuclear, and renewable plants built in 2015. The projections illustrate how different assumptions, such as for the availability of federal incentives, change the cost rankings of the technologies. Key observations include the following:

- Government incentives can change the relative costs of the generating technologies. For example, federal loan guarantees can turn nuclear power from a high cost technology to a relatively low cost option.
- The natural gas-fired combined cycle power plant, the most commonly built type of large natural gas plant, is a competitive generating technology under a wide variety of assumptions for fuel price, construction cost, government incentives, and carbon controls. This raises the possibility that power plant developers will continue to follow the pattern of the 1990s and rely heavily on natural gas plants to meet the need for new power generation.
- With current technology, coal-fired power plants using carbon capture equipment are an expensive source of electricity in a carbon control case. Other power sources, such as wind, nuclear, geothermal, and the natural gas combined cycle plant without capture technology, currently appear to be more economical.

None of the projections is intended to be a "most likely" case. Future uncertainties preclude firm forecasts. The value of this discussion is not as a source of point estimates of future power costs, but as a source of insight into the factors that can determine future outcomes, including factors that can be influenced by the Congress.

The main body of report is divided into the following sections:

- Types of generating technologies;
- Factors that drive power plant costs;
- Financial analysis methodology;
- Analysis of power project costs.

The report also includes the following appendixes:

- Appendix A presents power generation technology process diagrams and images.
- Appendixes B and C provide the data supporting the capital cost estimates used in the economic analysis. Appendix C also shows how operating costs and plant efficiencies were estimated for certain carbon control technologies.
- Appendix D presents the financial and operating assumptions used in the power cost estimates.
- Appendix E is a list of acronyms used in the report.

Chapter 2

TYPES OF GENERATING TECHNOLOGIES

The first part of this section describes how the characteristics of electricity demand influence power plant choice and operation. The next part describes the generating technologies analyzed in the report.

Electricity Demand and Power Plant Choice and Operation

Generation and Load

The demand for electricity ("load") faced by an electric power system varies moment to moment with changes in business and residential activity and the weather. Load begins growing in the morning as people waken, peaks in the early afternoon, and bottoms-out in the late evening and early morning. **Figure 1** is an illustrative daily load curve.

The daily load shape dictates how electric power systems are operated. As shown in Figure 1, there is a minimum demand for electricity that occurs throughout the day. This base level of demand is met with "baseload" generating units which have low variable operating costs.[2] Baseload units can also meet some of the demand above the base, and can reduce output when demand is unusually low. The units do this by "ramping" generation up and down to meet fluctuations in demand.

The greater part of the daily up and down swings in demand are met with "intermediate" units (also referred to as load-following or cycling units). These units can quickly change their output to match the change in demand (that is, they have a fast "ramp rate"). Load-following plants can also serve as "spinning reserve" units that are running but not putting power on the grid, and are immediately available to meet unanticipated increases in load or to back up other units that go off-line due to breakdowns.

The highest daily loads are met with peaking units. These units are typically the most expensive to operate, but can quickly startup and shutdown to meet brief peaks in demand. Peaking units also serve as spinning reserve, and as "quick start" units able to go from shutdown to full load in minutes. A peaking unit typically operates for only a few hundred hours a year.

Economic Dispatch and Heat Rate

The generating units available to meet system load are "dispatched" (put on-line) in order of lowest variable cost. This is referred to as the "economic dispatch" of a power system's plants.

For a plant that uses combustible fuels (such as coal or natural gas) a key driver of variable costs is the efficiency with which the plant converts fuel to electricity, as measured by the plant's "heat rate." This is the fuel input in British Thermal Units (btus) needed to produce one kilowatt-hour of electricity output. A lower heat rate equates with greater efficiency and lower variable costs. Other things (most importantly, fuel and environmental compliance costs) being equal, the lower a plant's heat rate, the higher it will stand in the economic dispatch priority order. Heat rates are inapplicable to plants that do not use combustible fuels, such as nuclear and non-biomass renewable plants.

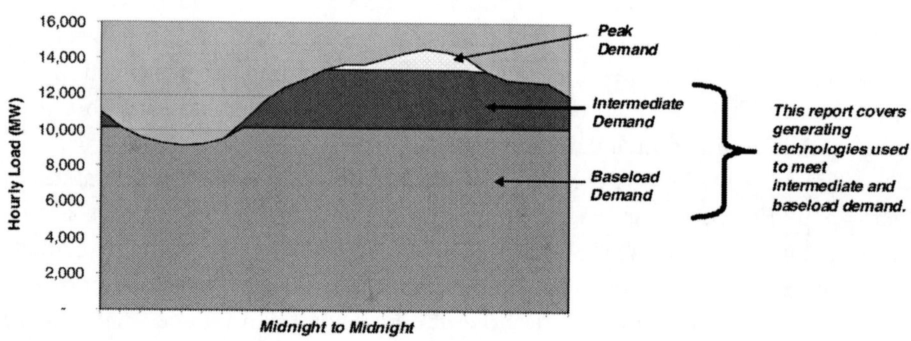

Figure 1. Illustrative Load Curve

As an illustration of economic dispatch, consider a utility system with coal, nuclear, geothermal, natural gas combined cycle, and natural gas peaking units in its system:

- Nuclear, coal, and geothermal baseload units, which are expensive to build but have low fuel costs and therefore low variable costs, will be the first units to be put on line. Other than for planned and forced maintenance, these baseload generators will run throughout the year.
- Combined cycle units, which are very efficient but use expensive natural gas as a fuel, will meet intermediate load. These cycling plants will ramp up and down during the day, and will be turned on and off dozens of times a year.
- Peaking plants, using combustion turbines,[3] are relatively inefficient and burn expensive natural gas. They run only as needed to meet the highest loads.[4]

An exception to this straightforward economic dispatch are "variable renewable" power plants — wind and solar — that do not fall neatly into the categories of baseload, intermediate, and peaking plants. Variable renewable generation is used as available to meet demand. Because these resources have very low variable costs they are ideally used to displace generation from gas-fired combined cycle plants and peaking units with higher variable costs. However, if wind or solar generation is available when demand is low (such as a weekend or, in the case of wind, in the evening), the renewable output could displace coal generation.

Power systems must meet all firm loads at all times, but variable renewable plants do not have firm levels of output because they are dependent on the weather. They are not firm resources because there is no guarantee that the plant can generate at a specific load level at a given point in time.[5] Variable renewable generation can be made firm by linking wind and solar plants to electricity storage, but with current technology, storage options are limited and expensive.[6]

Capacity Factor

As discussed above, baseload units run more often than cycling units, and peaking units operate the least often. The utilization of a generating unit is measured by its "capacity factor." This is the ratio of the amount of power generated by a unit for a period of time (typically a year) to the maximum amount of power the unit could have generated if it operated at full output, non-stop. For example, the maximum amount of power a 1,000 megawatt (MW) unit can generate in a year is 8.76 million megawatt-hours (Mwh), calculated as:

1,000 MW x 8,760 hours in a year = 8.76 million Mwh.

If this unit actually produced only 4.0 million Mwh its capacity factor would be 46% (calculated as 4.0 million Mwh divided by 8.76 million Mwh).

Note in this calculation the distinction between *capacity* and *energy*. Capacity is the potential instantaneous output of a generating unit, measured in watts.[7] Energy is the actual amount of electricity generated by a power plant during a time period, measured in watt-hours. The units are usually expressed in thousands (kilowatts and kilowatt-hours) or millions (megawatts and megawatt-hours).

The difference between actual and theoretical maximum output is caused by planned maintenance, mechanical breakdowns (forced outages), and any instances in which the plant is backed-down from maximum output due to lack of load or because the plant's power is more expensive than that from other plants. It is rare for a plant to have a capacity factor of 100%. Baseload plants typically have capacity factors of about 70% or greater, peaking plants about 25% or less, and cycling plants fall in the middle.

Utility Scale Generating Technologies

The types of generating technologies discussed in this report are often referred to as "utility scale" plants for baseload or intermediate service. These technologies generate large amounts of electricity at a single site for transmission to customers. In 2006, large baseload and intermediate service power plants accounted for about 86% of total power generation in the United States.[8] Utility scale plants typically have generating capacities ranging from dozens to over a thousand megawatts.

The one smaller scale generating technology covered in this report is solar photovoltaic power. The capacity of the largest U.S. central station solar photovoltaic plant, at Nellis Air Force Base in Nevada, is only 14 MW. Because of their small size, high capital costs, and low utilization rates, solar photovoltaic plants built with current technology have very high electricity production costs. Central station solar photovoltaic power is nonetheless included in the cost analysis because of public interest.

The report excludes peaking plants, which play an important but small role in the power system. The report also excludes oil-fired generation, which has all but disappeared from the nation's generating mix because of the high cost of the fuel. In 1978, oil-fired plants produced 22% of the nation's electricity. By 2007 the oil¬fired share was less than 2%.[9] Significant construction of new oil-fired plants is not expected.

The report also does not cover combined heat and power (CHP) plants. These are typically industrial plants that co-produce electricity and steam for internal use and for sale. Unlike plants that generate power exclusively to put electricity on the grid, CHP facilities have unique, plant-specific operating modes and cost structures, and economics fundamentally different from utility scale generation. CHP generation is a small part of the electric power industry, accounting for about 3.7% of total electricity output in 2007.[10] Hydropower is excluded because no significant construction of new, large hydroelectric plants is expected (due to environmental concerns and the small number of available sites).[11]

The cost analysis is for plants entering service on January 1, 2015, which means construction would start soon (between 2009 and 2013 depending on the technology). The plants therefore incorporate only small projected changes from 2008 cost and performance for mature technologies, and reflect current estimates of cost and performance for new or evolving technologies (such as advanced nuclear power and coal gasification).

The technologies covered in the report are described briefly below. Process diagrams and images of each technology are in Appendix A.

Supercritical Pulverized Coal

Pulverized coal plants account for the great majority of existing and planned coal-fired generating capacity. In this system coal is ground to fine power and injected with air into a boiler where it ignites. Combustion heat is absorbed by water-carrying tubes embedded in the boiler walls and downstream of the boiler. The heat turns the water to steam, which is used to rotate a turbine and produce electricity. Since about 2000 most plans for new pulverized coal plants have been for "supercritical" designs that gain efficiency by operating at very high steam temperatures and pressures.

In 2007, coal generation of all types[12] accounted for 49% of total power generation in the United States (see **Figure 2**).

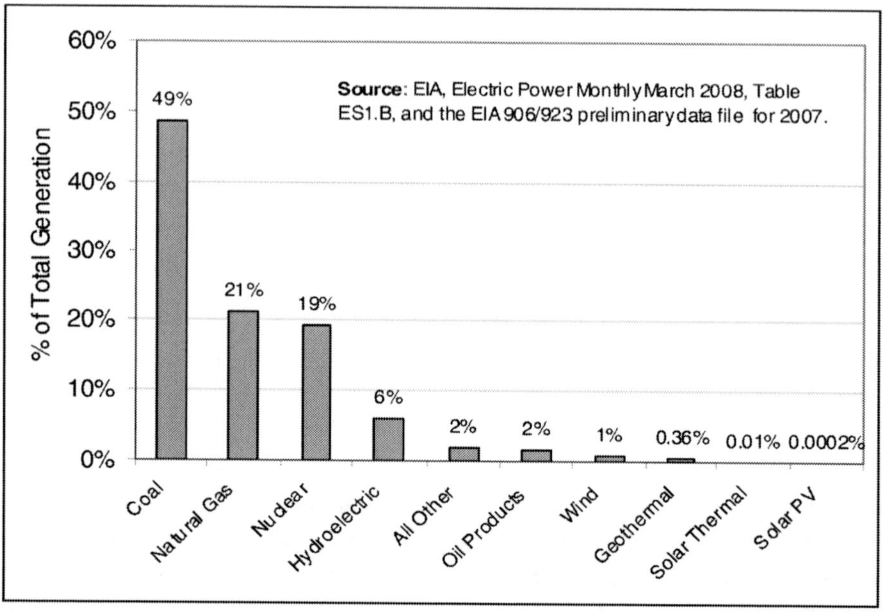

Figure 2. Total U.S. Electric Power Generation by Energy Source, 2007

Integrated Gasification Combined Cycle (IGCC)

In this process coal is converted to a "synthesis gas" (syngas) before combustion. IGCC plants are more expensive to build than pulverized coal generation, but proponents believe they have compensating advantages, including:

- Lower emissions of air pollutants, such as sulfur dioxide (SO_2), nitrogen oxides (NOx), and mercury. However, modern pulverized coal plants also have low emissions of air pollutants, so the advantage of IGCC plants over conventional technology is limited.
- Greater efficiency (i.e., a lower heat rate), although with current technology IGCC has only a small efficiency advantage over conventional coal plants.[13]
- The syngas that results from the gasification process can be processed to convert the carbon in the gas into a concentrated stream of carbon dioxide (CO_2). The syngas can then be processed, before it is burned, to remove the CO_2.

In principle this pre-combustion capture of CO_2 can be accomplished more easily and cheaply than post-combustion removal of CO_2 from the exhaust gases ("flue gas") emitted by a conventional coal plant. The promise of more efficient carbon capture is one of the primary rationales for IGCC technology.

Coal-fired IGCC experience in the United States is limited to a handful of research and prototype plants, none of which is designed for carbon capture. A commercial IGCC plant is being constructed by Duke Energy at its Edwardsport site in Indiana, and other projects have been proposed. However, some other power plant developers will not build IGCC plants because of concerns over cost and the reliability of the technology.[14] In general, the cost and operational advantages of IGCC over conventional coal technology and the commercial readiness of IGCC technology are disputed.[15]

Natural Gas Combined Cycle

Combined cycle plants are built around one or more combustion turbines, essentially the same technology used in jet engines. The combustion turbine is fired by natural gas to rotate a turbine and produce electricity. The hot exhaust gases from the combustion turbine are captured and used to produce steam, which drives another generator to produce more electricity. By converting the waste heat from the combustion turbine into useful electricity the combined cycle achieves very high efficiencies, with heat rates below 7,000 btus per kWh (compared to around 9,000 btus per kWh for new pulverized coal plants). This high efficiency partly compensates for the high cost of the natural gas used in these plants.

Modern combined cycle plants, which evolved in the 1990s, have a relatively low construction cost and modest environmental impacts; can be used to meet baseload, intermediate, and peaking demand; can be built quickly; and are very efficient. Because of these advantages, since 1995 natural gas combined cycle plants have accounted for 88% of the all the new generating capacity built in the United States capable of baseload and intermediate service.[16]

Natural gas combined cycle plants and other types of gas-fired power plants are expected to continue to dominate capacity additions into the next decade.[17] According to EIA, combined cycle plants will account for 29% of all capacity additions between 2008 and 2015.[18] However, this forecast may understate actual combined cycle plant additions. The EIA estimates that coal plants will account for almost a quarter of new capacity built through 2015, the equivalent of about 170 new coal-fired generating units.[19] It is questionable whether this much coal capacity will actually be built because of public opposition to new coal plants and the cost of the plants. Utilities reportedly canceled 16,577 MW of planned generating capacity in 2007, of which 84% was coal-fired.[20] According to a Department of Energy (DOE) report, only 12% (4,500 MW) of the coal capacity planned in 2002 to be built by 2007 was actually constructed. The report notes that "delays and cancellations have been attributed to regulatory uncertainty (regarding climate change) or strained project economics due to escalating costs in the industry."[21]

If less coal capacity is built than planned, the main replacement is likely to be combined cycle plants, the type of gas-fired unit capable of replacing a baseload coal plant. For example, in 2007, power generators in Florida planned to install 4,627 MW of new coal fired capacity through 2016. By 2008 the plans for new coal-fired capacity had dropped to 738 MW, primarily "due to environmental concerns at the State level. The majority of this decrease in planned coal-fired generation was replaced with gas-fired units."[22]

Natural gas combined cycle plants accounted for 17% of total generation in 2007,[23] and natural gas plants of all types accounted for 21% of total power generation in the United States (Figure 2).

Nuclear Power

Nuclear power plants use the heat produced by nuclear fission to produce steam. The steam drives a turbine to generate electricity. Nuclear plants are characterized by high investment costs but low variable operating costs, including low fuel expense. Because of the low variable costs and design factors, nuclear plants in the United States operate exclusively as baseload plants and are typically the first plants in a power system's dispatch order. Nuclear power supplied 19% of the nation's electricity in 2007 (Figure 2).

This report discusses projected costs for Generation III/III+ technology nuclear plants. These plants are more advanced versions of the 104 reactors currently operating in the United States, and all reactors currently proposed for construction in the United States are Generation III/III+ designs. Compared to existing reactors, the Gen III/III+ plants are designed to reduce costs and enhance safety through, for example, reduced complexity, standardized designs, and improved construction techniques. Some designs also incorporate passive safety systems that are supposed to be capable of preventing a catastrophic accident even without operator action.

There are several competing Gen III/III+ designs,[24] but only one design has been built (General Electric's Advanced Boiling Water Reactor, of which four units have been constructed in Japan). Plants based on other Gen III/III+

designs are under construction in France, Finland, and China. As discussed later in the report, the costs of building a new nuclear plant in the United States will apparently be very high.

Geothermal Power

Geothermal plants have operated for many years in the western United States, mainly in California. In a typical binary cycle geothermal facility, wells draw hot water and steam from underground into a heat exchanger. In the heat exchanger a working fluid is vaporized and used to drive a turbine generator (the underground steam is not used directly because it contains corrosive impurities and can release air pollutants). In geothermal fields that have been depleted by years of use, such as the Geysers field in California, operators can inject water into the layers of hot rock to supplement the naturally available water and boost steam production. Unlike solar and wind power, which are weather-dependent, geothermal plants operate as dispatchable baseload plants. However, with current technology, geothermal plants are limited to small facilities (typically under 50 MW) at sites in the western United States.[25] In 2007, geothermal plants produced 0.4% of the nation's power supply (Figure 2).[26]

Wind Power

Wind power plants (sometimes referred to as wind farms) use wind-driven turbines to generate electricity. An individual turbine typically has a capacity in the range of 1.5 to 2.5 MW, and a wind plant installs dozens or hundreds of these turbines. As noted above, wind is a variable renewable resource because its availability depends on the vagaries of the weather. Wind supplied 1% of total U.S. power supply in 2007 (Figure 2); EIA estimates that assuming no changes to current law and regulation, this will increase to 2.4% by 2030.[27]

Solar Thermal and Solar Photovoltaic (PV) Power

Solar thermal and PV power are alternative means of harnessing sunlight to produce electricity. PV power uses solar cells to directly convert sunlight to electricity. To date most of the solar PV installations in the United States have been small (about one MW or less). Two exceptions are the installations at Nellis Air Force Base in Nevada (14 MW) and the Alamosa Photovoltaic Power Plant in Colorado (8 MW).

Solar thermal plants, also referred to as concentrated solar power (CSP), concentrate sunlight to heat a working liquid to produce steam that drives a power- generating turbine. Two major types of solar thermal systems are parabolic trough and power tower technologies. Parabolic trough plants use an array of mirrors to focus sunlight on liquid-carrying tubes integrated with the mirrors. Several parabolic trough installations have operated successfully in California since the 1 980s, and the 64 MW Nevada Solar One plant began operating in 2007.

The power tower technology uses a mirror field to focus sunlight on a central tower, where the heat is used to produce steam for power generation. A research power tower, the Solar One/Two plant, operated for several years in the 1980s and 1 990s in California. A power tower plant has recently been constructed in Spain and a 400 MW project has been proposed for California.

Several new solar thermal projects, primarily of the parabolic trough and related types, are in development. The capacity of these projects range up to 554 MW. A potential advantage of solar thermal systems is the ability to produce electricity when sunlight is weak or unavailable by storing solar heat in the form of molten salt. If storage proves economical for large-scale plants, then solar thermal facilities in regions with strong, near continuous daytime sunlight, such as the Mojave desert, could be operated as dispatchable plants with firm capacity.

In 2007, solar thermal generation accounted for 0.0 1% of total generation, and solar PV power for less (**Figure 2**).

Chapter 3

FACTORS THAT DRIVE POWER PLANT COSTS

This section of the report discusses the major factors that determine the costs of building and operating power plants. These factors include:

- Government incentives.
- Capital (investment) cost, including construction costs and financing.
- Fuel costs.
- Air emissions controls for coal and natural gas plants.

Government Incentives

Many government incentives influence the cost of generating electricity. In some cases the incentives have a direct and clear influence on the cost of building or operating a power plant, such as the renewable investment tax credit. Other programs have less direct affects that are difficult to measure, such as parts of the tax code that influence the cost of producing fossil fuel.[28]

The economic analysis in this report incorporates the following incentives that directly affect the cost of building or operating power plants.[29]

Renewable Energy Production Tax Credit[30]

The credit has a 2008 value of 2.0 cents per kWh, with the value indexed to inflation. The credit applies to the first 10 years of a plant's operation. As of October 2008 the credit is available to plants that enter service before the end of 2009. The credit is currently available to new wind, geothermal, and several other renewable energy sources. New solar energy projects do not qualify, and geothermal projects can take the production tax credit only if they do not use the renewable investment tax credit (discussed below).

Nuclear Energy Production Tax Credit[31]

The credit, which is for new advanced nuclear plants, has a nominal value of 1.8 cents per kWh. The credit applies to the first eight years of plant operation. Unlike the renewable production tax credit the nuclear credit is not indexed to inflation and therefore drops in real value over time. This credit is subject to several limitations:

- It is available to advanced (i.e., Gen III/III+) nuclear plants that begin construction before January 1, 2014, and enter service before January 1, 2021.
- For each project the annual credit is limited to $125 million per thousand megawatts of generating capacity.
- The full amount of the credit will be available to qualifying facilities only if the total capacity of the qualifying facilities is 6,000 megawatts or less. If the total qualifying capacity exceeds 6,000 megawatts the amount of the credit available to each plant will be prorated. EIA estimates in its 2008 *Annual Energy Outlook* that 8,000 megawatts of new nuclear capacity will qualify;[32] in this case the credit amount would drop to 1.35 cents per kWh once all the qualifying plants are on-line. This pro-rated value is used in the report's economic analysis of generating costs.

Loan Guarantees for Nuclear and Other Carbon-Control Technologies[33]

Under final Department of Energy (DOE) rules the loan guarantees can cover up to 80% of the cost of a project, and are awarded based on a detailed evaluation of each applicant project. Entities receiving loan guarantees must make a "credit subsidy cost" payment to the federal treasury that reflects the anticipated cost of the guarantee to the government, including a probability weighted cost of default. Because the debt is backed by the federal government, it is expected to carry the highest credit rating and therefore a low interest rate.[34] The guarantees are unavailable to publicly owned utilities, such as municipal systems.[35]

Congress periodically determines the total value of the guarantees that the DOE is authorized to grant. In April 2008, the Department of Energy announced plans to solicit up to $18.5 billion in loan guarantee applications for nuclear projects.[36] As of November 2008, DOE was considering several applications for loan guarantees.

Developers and investors have stated that the loan guarantees are critical to constructing at least the first wave of new nuclear plants. This is because of the multi-billion dollar cost of a nuclear project, which can exceed the total market value of the company building a plant. For example, in 2008 the president of Exelon Generation, which operates a large fleet of existing nuclear plants and plans to build new units, stated that constructing new nuclear plants would be "impossible" without loan guarantees.[37]

Energy Investment Tax Credit[38]

Tax credits under this program are available to solar and geothermal electricity generation, and some other innovative energy technologies. Wind energy systems do not qualify. The credit is 10% for geothermal systems, and is 30% for solar electric systems installed before January 1, 2017 (after which it reverts to 10%). Geothermal projects that take the investment tax credit cannot claim the renewable production tax credit.[39] The depreciable basis of the project for tax purposes is reduced by 50% of the credit value. The investment tax credit is available to independent power producers and investor owned utilities, but is inapplicable to tax-exempt publicly owned utilities.[40]

Clean Coal Technologies Investment Tax Credit[41]

This tax credit can be used by investor owned utilities or independent power producers (it is inapplicable to tax-exempt publicly owned utilities). It is limited to a total of $2.55 billion in tax credits, of which (1) $0.8 billion is specifically for IGCC plants; (2) $0.5 billion is for non-IGCC advanced coal technologies, and (3) $1.25 billion is for advanced coal projects generally. The tax credits in the third category will not be awarded until after the program that encompasses the first two categories of tax credits is completed or until such other date designated by the Secretary of Energy.[42] The depreciable basis of a project for tax purposes is reduced by 50% of the credit value.

State and Local Incentives

State and local governments can offer additional incentives, such as property tax deferrals. The combined value of the government tax breaks can run into the hundreds of millions of dollars per project. For example, Duke Energy's Edwardsport IGCC project in Indiana is expected to receive almost half-a-billion dollars in federal, state, and local tax incentives.[43]

State utility commissions can use rate treatment of new plants as a financial incentive for the investor owned utilities they regulate. Under traditional rate making a utility is not permitted to earn a return on its construction investment until a plant is in service. This approach to ratemaking is used to motivate the utility to prudently manage construction, and to ensure that customers do not have to pay for a power plant until it is operating. However, if a project is very expensive, the time lag between when costs are incurred and when return on the investment is allowed in rates can put a financial strain on the company. If the plant is expensive, adding the return into rates as a single big adjustment can inflict "rate shock" on customers.

For these reasons, utilities sometimes argue for an alternative rate making method called "construction work in progress (CWIP) in rates." In this approach, a utility is allowed to recover in rates the return on its investment as the plant is being built. CWIP in rates relieves the utility of the financial strain of carrying an expensive investment that is yielding no income, phases-in the rate increase to customers, and decreases the utility's financial exposure if the project is delayed. On the other hand, the pressures for prudent construction management inherent in traditional ratemaking are dampened.

Some states, such as South Carolina and Mississippi, have passed legislation allowing utility projects that meet certain criteria to receive CWIP in rates.[44] In other cases utilities have received CWIP in rates under existing rules. CWIP in rates has expanded beyond its historic application to very expensive coal and nuclear projects. For example, the Kansas and Wisconsin commissions have allowed CWIP in rates for relatively small wind projects.[45]

Capital and Financing Costs

Construction Cost Components and Trends

Most of the generating technologies discussed in this report are capital intensive; that is, they require a large initial construction investment relative to the amount of generating capacity built. Power plant capital costs are often discussed in terms of dollars per kilowatt (kW) of generating capacity. All of the technologies considered in this report have estimated 2008 costs of $2,100 per kW or greater, with the exception of the natural gas combined cycle plant ($1,200 see Appendix B). Nuclear, geothermal, and IGCC plants have estimated costs in excess of $3,000 per kW.

Power plant capital costs have several components. Published information on plant costs often do not clearly distinguish which components are included in an estimate, or different analysts may use different definitions. The capital cost components are:

- Engineering, Procurement, and Construction (EPC) cost: this is the cost of the primary contract for building the plant. It includes the cost of designing the facility, buying the equipment and materials, and construction.[46]
- Owner's costs: these are any construction costs that the owner handles outside the EPC contract. This could include arranging for the construction of transmission and fuel delivery facilities (such as a natural gas pipeline) to a power plant.
- Capitalized financing charges: a plant developer incurs financing charges while a power plant is being built. This includes interest on debt and an imputed cost of equity capital. Until the plant is operating these costs are capitalized; that is, become part of the investment cost of the project for tax, regulatory, and financial analysis purposes (see further discussion of financing costs, below).

Construction costs for power plants have escalated at an extraordinary rate since the beginning of this decade. According to one analysis, the cost of building a power plant increased by 131% between 2000 and 2008 (or by 82% if nuclear plants are excluded from the estimate). Costs reportedly increased by 69% just since 2005. The cost increases affected all types of generation. For example, between 2000 and 2008, the cost of wind capacity reportedly increased by 108%, coal increased by 78%, and gas-fired plants by 92%.[47] The cost increases have been attributed to many factors, including:

- High prices for raw and semi-finished materials, such as iron ore, steel, and cement.
- Strong worldwide demand for generating equipment. China, for example, is reportedly building an average of about one coal-fired generating station a week.[48]
- Low value of the dollar.
- Rising construction labor costs, and a shortage of skilled and experienced engineering staff.[49]
- An atrophied domestic and international industrial and specialized labor base for nuclear plant construction and components.

- In the case of wind, competition for the best plant sites and a tight market for wind turbines; in the case of nuclear plants, limited global capacity to produce large and ultra-large forgings for reactor pressure vessels.[50]
- Coincident worldwide demand for similar resources from other business sectors, including general construction and the construction of process plants such as refineries. Much of the demand is driven by the rapidly growing economies of Asia.[51]

The future trend in construction costs is a critical question for the power industry. Continued increases in capital costs would favor building natural gas plants, which have lower capital costs than most alternatives. Stable or declining construction costs would improve the economics of capital-intensive generating technologies, such as nuclear power and wind.[52] At least some long-term moderation in cost escalation is likely, as demand growth slackens and new supply capacity is added.[53] But when and to what degree cost increases will moderate is as unpredictable as the recent cost escalation was unforeseen.

Financing Power Plant Projects

Even relatively small power plants cost millions of dollars. For example, the capital cost for a 50 MW wind plant would be about $105 million at $2,100 per kW of capacity. The investment cost is typically financed by a combination of debt and equity.[54] The financing structure and the cost of money depends on the type of developer and project-specific risk.

Three types of entities typically develop power plants:

- **Investor-owned utilities (IOUs):** IOUs are owned by private investors and are subject to government regulation of rates and conditions of service. They have guaranteed service territories and face limited competition. State utility commissions set electric rates designed to maintain the financial health of the utility, assuming it operates prudently. The commission also must approve proposals by the utility to build new power plants. [55]

- **Publicly-owned utilities (POUs):** A POU is a utility that is an agency of a municipality, a state, or the federal government. Electric cooperatives are also considered to be POUs. Like IOUs, POUs have guaranteed service territories and face limited competition. Most POUs are small, provide only distribution service, and have limited financial and management resources.[56] But larger and some smaller POUs also own and operate power plants, sometimes as co- owners of projects where an IOU or independent power producer is the lead developer. Examples of POUs with large amounts of generation include the Tennessee Valley Authority and the municipal utilities serving the cities of Los Angeles and San Antonio. POUs set their own rates and make their own decisions to build power plants.

- **Independent Power Producers (IPPs):** IPPs are merchant developers and operators of power plants that sell wholesale power to utility and industrial buyers. Within limits they can sell power at whatever price the market will bear.[57] IPPs face more financial risk than regulated utilities — they do not have guaranteed service territories and can face intense competition for power sales — but can also earn larger profits. IPPs make their own decisions to build power plants.

All three types of entities play a major role in the electric power industry (**Table 1**). The lines between the entities can blur. Holding companies that own IOUs can also own IPPs. POUs sometimes own large shares of power projects developed by IOU or IPPs.

Table 1. Shares of Total National Electric Generation and Generating Capacity, 2006

	Generation	Generating Capacity
Publicly-Owned Utilities	22%	21%
Investor-Owned Utilities	41%	38%
Non-Utilities	37%	41%
National Total	100%	100%

Source: American Public Power Association [http://www.appanet.org/files /PDFs/nameplate2006.pdf], citing Energy Information Administration.

Notes: Non-utility generation includes independent power producers and power marketers. Non-utility capacity includes industrial and commercial facilities. Capacity shares are for nameplate capacity.

The cost of the money used to finance power projects varies significantly between IOU, POUs, and IPPs. A POU will normally finance a project with 100% debt at a low interest rate. The rate is low because interest paid on public debt is exempt from federal or state income taxes,[58] and because public entities have a very low risk of default (failure to make debt payments), much lower than for private businesses.[59] Typical municipal bonds have ratings in the middle or upper tiers of investment grade debt.[60]

Privately owned IOUs and IPPs finance power projects with a mix of debt and equity. Debt is more costly to these companies than to POUs because it is not tax exempt and because they usually have lower credit ratings. The electric utility industry as a whole has a credit rating in the lower tier of the investment grade category (BBB).[61] IPP debt often falls in the speculative category and has a higher interest rate than IOU or POU issues.[62]

Investors expect private developers to make a significant equity contribution to a project.[63] Reliance on equity versus debt varies by company and project. The cost analysis used in this study assumes that IPPs and IOUs rely on, respectively, 40% and 50% equity (see **Table 17** in Appendix D), except in the case where federal loan guarantees are available (see discussion of government incentives, above). Equity is more expensive than debt,[64] and is more expensive for IPPs than IOUs because IPPs typically face more competition and financial risk.

In summary:

- Because POUs can finance a power project with 100% low-cost debt they can build power plants more cheaply than IOUs or IPPs. However, because of the small size of most POUs they do not have the financial or management resources to take on large and complex projects by themselves, so POUs often partner on projects where an IOU or IPP is the lead developer.
- IOU's typically have lower financing costs than IPP's because they have lower costs of debt and equity.[65]
- Financing costs are highest for IPPs, which makes them somewhat less prone to take on the highest cost projects (such as coal and nuclear plants) unless POUs or IOUs are co-owners.

Fuel Costs

Fuel costs are important to the economics of coal, nuclear, and natural gas plants, and irrelevant to solar, geothermal, and wind power. Recent trends in the delivered cost of coal and natural gas to power plants are illustrated below in **Figure 3.** The constant dollar prices of both fuels have increased since the beginning of the decade, but the price escalation has been especially severe for natural gas.[66] Natural gas has also been consistently more expensive than coal. The comparatively low cost of coal partly compensates for the high cost of building coal plants, while the high cost of natural gas negates part of the capital cost and efficiency advantages of combined cycle technology.

Because it takes years to build a power plant, and plants are designed to operate for decades, generation plans largely pivot on fuel price forecasts. However, fuel prices have been notoriously difficult to predict. For example, EIA forecasts of delivered coal prices and natural gas wellhead prices have been off target by an average of, respectively, 47% and 64%.[67] EIA attributes the gap between actual and forecasted gas prices to a host of factors:

> As regulatory reforms that increased the role of competitive markets were implemented in the mid-1980s, the behavior of natural gas was especially difficult to predict. The technological improvement expectations embedded in early AEOs [Annual Energy Outlooks] proved conservative and advances that made petroleum and natural gas less costly to produce were missed. After natural gas curtailments that artificially constrained natural gas use were eased in the mid-1980s, natural gas was an increasingly attractive fuel source, particularly for electricity generation and industrial uses. Historically, natural gas price instability was strongly influenced by natural gas resource estimates, which steadily rose, and by the world oil price. More recently, the AEO reference case has overestimated natural gas consumption due to the use of natural gas wellhead price projections that proved to be significantly lower than what actually occurred.[68]

EIA's analysis illustrates how the confluence of technological, regulatory, resource, and domestic and international market factors make fuel forecasts so problematic. Fuel price uncertainty is especially important in evaluating the economics of natural gas-fired combined cycle plants. For the base assumptions used in this study, fuel constitutes half of the total cost of power from a new combined cycle plant, compared to 18% for a coal plant and 6% for a nuclear plant.

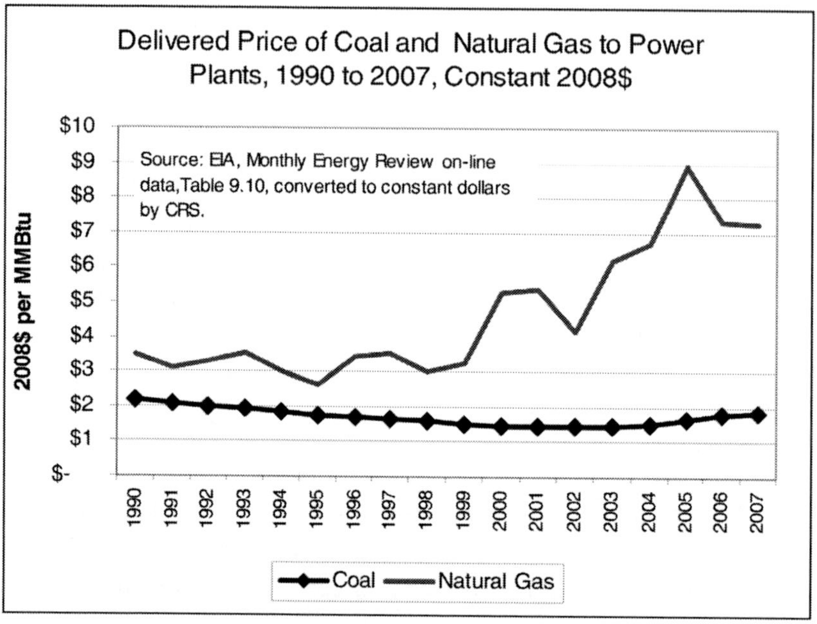

Figure 3. Coal and Natural Gas Constant Dollar Price Trends

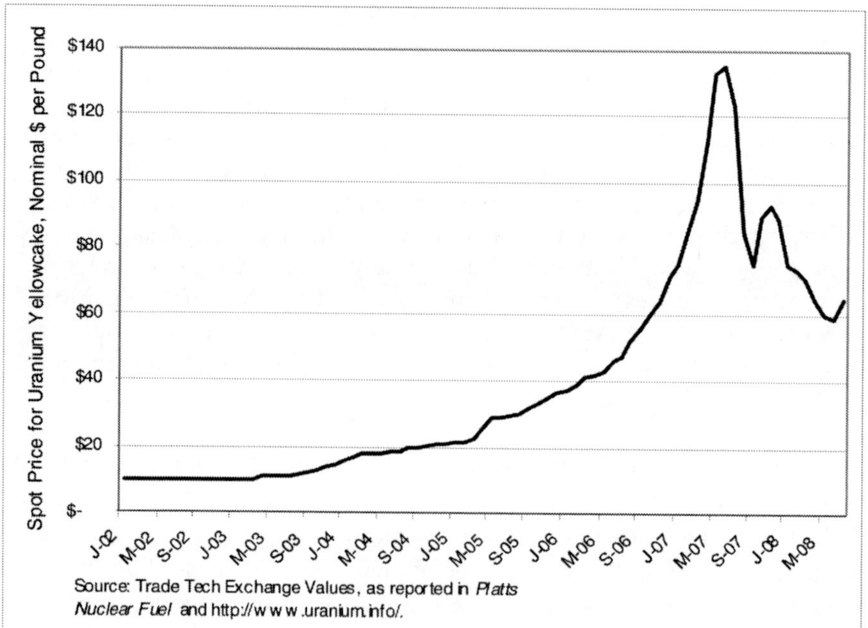

Figure 4. Uranium Price Trends

The price of the uranium used to make nuclear fuel has, like coal and natural gas, increased sharply and has been volatile (**Figure 4**). Although prices have recently dropped, they are still far above historic levels.[69] Over the long term, EIA expects nuclear fuel prices to increase in real terms from $0.58 per mmbtu in 2007 to $0.77 per mmbtu in 2023, and then slowly decline.[70] Even prices twice as high would not have a major impact on nuclear plant economics, which are dominated by the capital cost of building the plant.

Air Emissions Controls for Coal and Gas Plants

Regulations that limit air emissions from coal and natural gas plants can impose two types of costs: The cost of installing and operating control equipment, and the cost of allowances[71] that permit plants to emit pollutants. The following emissions are discussed below:

Emissions from coal:

- Sulfur dioxide (SO_2), a precursor to acid rain and the formation in the atmosphere of secondary particulates[72] that are unhealthy to breathe and can impair visibility.
- Mercury, a toxic heavy metal.
- Primary particulates (soot) entrained in the power plant's flue gas.

Emissions from coal and natural gas:

- Nitrogen oxides (NOx), a precursor to ground level ozone, acid rain, and the formation in the atmosphere of secondary particulates.
- Carbon dioxide (CO_2), a greenhouse gas produced by the combustion of fossil fuels.

The regulations and control technologies for SO_2, NOx, particulates, and mercury are discussed briefly under the category of "conventional emissions." These pollutants are subject to either existing regulations or regulations being developed under current law, and can be controlled with well-understood, commercially- available technologies. CO_2 is discussed in more detail because control technologies are still under development and may be far more costly than controls for conventional emissions.[73] While CO_2 is not currently subject to federal regulation, control legislation is being actively considered by the Congress and some states are taking action to limit CO_2 emissions.

More information on air emissions, particularly on regulatory and policy issues, is available in numerous CRS reports. The reports can be accessed through the "Energy, Environment, and Resources" link on the CRS website, [http://www.crs.gov].

Conventional Emissions

The Environmental Protection Agency (EPA) has established National Ambient Air Quality Standards (NAAQS) for several pollutants, including SO_2, NOx, ozone, and particulates. New coal and natural gas plants built in areas in compliance with a NAAQS standard must install Best Available Control Technology (BACT) pollution control equipment that will keep emissions sufficiently low that the area will stay in compliance. Plants built in areas not in compliance with a NAAQS (referred to as "non-attainment" areas) must meet a tighter Lowest Achievable Emission Rate (LAER) standard.[74] In practice, air permit emissions are negotiated case-by-case between the developer and state air authorities. Federal standards set a ceiling; state permits can specify lower emission limits.

In addition to technology control costs, new plants that emit SO_2 must buy SO_2 emission allowances under the acid rain control program established by Title IV of the Clean Air Act.[75] Depending on the location of a new plant, it may also need to purchase NOx allowances.[76]

Regulation of mercury is unsettled. On February 8, 2008, the U.S. Court of Appeals for the D.C. Circuit vacated the Bush administration's Clean Air Mercury Rule, which would have allowed new coal plants to comply with mercury emission limits by purchasing mercury allowances. Because of the court's action, coal plant mercury emissions are now categorized as a hazardous air pollutant. If the decision stands,[77] it will trigger a requirement for all coal plants, old and new, to install mercury control equipment that meets a Maximum Available Control Technology (MACT) standard. EPA has not yet defined a MACT standard for mercury, but state air officials will probably require new plants to meet tight mercury emission limits.[78]

The technology and costs for controlling sulfur, NOx, particulate, and mercury emissions are briefly described below. For additional information on emission control technologies see the International Energy Agency Clean Coal Center at [http://www.iea-coal.org/site/ieacoal/databases/clean-coal-technologies].

- **Sulfur.** Commercial technologies can remove 95% to 99% of the SO_2 formed by burning coal in pulverized coal plants, and over 99% of the sulfur in IGCC synthesis gas before it is burned. To the degree that a new pulverized coal unit or IGCC plant releases SO_2 to the atmosphere, it must buy SO_2 emission allowances. Because SO_2 emissions by plants with controls are so small, allowances are not a major expense compared to the other costs of running a power plant. At mid-2008 allowance and fuel prices, the annual cost of SO_2 allowances for a coal plant burning eastern coal would be on the order of $1 million, compared to over $220 million just for fuel.[79] The cost of the control equipment is more significant. An SO_2 control system will account for about 12% of the capital cost of a new pulverized coal plant and 29% of non-fuel operating costs (**Table 2**). (It is difficult to isolate environmental control costs for an IGCC plant because emissions control is largely integral with cleanup of the synthesis gas that is necessary, irrespective of environmental rules, prior to combustion.)

- **Mercury.** Some pulverized coal plants can achieve 90% removal of mercury as a co-benefit of operating SO_2 and particulate control equipment. Other plants will have to install a powdered activated carbon injection system (accounting for about 1% of the plant's capital cost and 9% of non-fuel operating costs). IGCC plants would remove 90% to 95% of the mercury from the synthesis gas using another technology also based on activated carbon.

- **NOx.** Commercial technologies can reduce NOx emissions to very low levels for pulverized coal and IGCC plants. Depending on a plant's location, it may have to purchase NOx emission allowances. As in the case of SO_2 allowances, because the controlled emission rates for new plants are so low the total cost of allowances is small compared to other plant operating costs. The cost of the control equipment for a pulverized coal plant is about 2% of capital expense and 9% of non-fuel operating costs.

- **Particulates.** Primary particulates are controlled using removal systems that have been a standard feature of pulverized coal plants for many years. Removal efficiencies exceed 99%. Primary particulate removal rates for IGCC plants are expected to be similar. Secondary particulates are controlled by reducing NOx and SO_2 emissions, as discussed above.

Carbon Dioxide

This section of the report discusses the technical and cost characteristics of carbon control technologies for coal and natural gas plants. The estimates of the cost and performance affects of installing carbon controls are uncertain because no power plants have been built with full-scale carbon capture. For additional information on carbon control technologies, see CRS Report RL3 4621, *Capturing CO_2 from Coal-Fired Power Plants: Challenges for a Comprehensive Strategy*, by Larry Parker, Peter Folger, and Deborah D. Stine; and Steve Blankinship, "The Evolution of Carbon Capture Technology, Parts 1 and 2," *Power Engineering*, March and May 2008.[80]

Table 2. Emission Controls as an Estimated Percentage of Total Costs for a New Pulverized Coal Plant

	Percent of Total Cost	
	Plant Capital Cost	Plant O&M Cost
SO_2 Controls	12%	29%
NO_x Controls	2%	12%
Mercury Controls	1%	9%
Total for Emission Controls	16%	51%

Source: Calculated by CRS from MIT, *The Future of Coal, 2007*, Tables A-3.D.3. and Tables A-3.D.4. Calculations were made for the point estimates in the report; the tables have cost ranges for capital costs and for mercury control O&M costs.

Notes: SO_2 = sulfur dioxide; NOx = nitrogen oxides; O&M = operations and maintenance.

CO_2 Removal for Pulverized Coal and Natural Gas Plants

Technology developed by the petrochemical industry, using a class of chemicals called amines, can be used to scrub CO_2 from flue gas. Amine scrubbing is currently used to extract CO_2 from part of the flue gas at a handful of coal-fired plants, to produce CO_2 for enhanced oil recovery and the food industry, but the scale is about a tenth of what would be needed to scrub 90% of the CO_2 from the entire flue gas stream of a large power plant.[81] Scaling up amine technology to handle much larger gas flows at a power plant may be technically challenging.

Amine scrubbing is energy intensive. It diverts steam from power production and uses part of the plant's electricity production to compress the CO_2 for pipeline transportation to its final disposition. Amine scrubbing is estimated to cut a coal plant's electricity output by about 30% to 40%.[82] The equipment is also costly. According to one study, the cost for building a new coal plant with amine scrubbing is an estimated 61% higher than building the a plant without carbon controls.[83] The same study estimated the cost for a coal plant retrofit installation, without taking into account the recent rapid increase in power plant construction costs, at about $1,600 per kW of net capacity, or almost $1 billion for a 600 MW plant.[84]

The cost and performance impacts for adding amine scrubbing to a natural gas- fired combined cycle are also large. The estimated reduction in net electricity output is 14%, and the estimated increase in the plant capital cost is about 100%.[85] Researchers are attempting to commercialize less costly carbon capture technologies for conventional coal and gas plants, but these are still in early development.

Table 3. Estimates of the Change in IGCC Plant Capacity and Capital Cost from Adding Carbon Capture

Source and IGCC Technology	Change in Net Generating Capacity	Change in Plant Cost
NETL, 2007		
GE/Radiant	-13%	32%
CoP E-Gas	-17%	40%
Shell	-19%	35%
EIA, 2008		
Generic	n/a	43%
EPRI 2006		
Shell	-25%	51%
MIT 2007		
GE/Full Quench (retrofit)	-17%	n/a
CoP E-Gas (retrofit)	-28%	n/a
Generic	-28%	32%

Sources: NETL, *Cost and Performance Baseline for Fossil Energy Plants, Volume 1*, Exhibit 3-114; EIA, *Assumptions to the Annual Energy Outlook 2008*, Table 38; EPRI, *Feasibility Study for an Integrated Gasification Combined Cycle Facility at a Texas Site*, October 2006, Tables 7-1, 13-2, and 13-3; MIT, *The Future of Coal*, 2007, pp. 122, 150, and 151, and Table 30.

Notes: IGCC = Integrated Gasification Combined Cycle; NETL = National Energy Technology Laboratory; EIA = Energy Information Administration; EPRI = Electric Power Research Institute; MIT = Massachusetts Institute of Technology; n/a = not available; GE = General Electric; CoP = ConocoPhillips. Radiant and full quench refer to alternative means of heat capture from cooling of the synthesis gas. Values are for units built to incorporate carbon capture, except when retrofit is indicated.

CO_2 Removal for IGCC Coal Plants

Carbon capture for an IGCC plant involves multi-step treatment of the synthesis gas using technology originally developed for the petrochemical industry. Estimates of the cost and performance impact of incorporating carbon capture into a IGCC design vary widely. For the sample of studies shown in **Table 3**, the estimated increase in capital costs ranges from 32% to 51%. The estimated loss in generating capacity varies by more than a factor of two, from 13% to 28%. This wide variation reflects in part factors specific to different IGCC technologies, but is also an indication of limited experience with IGCC technology generally and the integration of carbon capture in particular.

While IGCC technology is arguably better-suited for carbon capture than pulverized coal systems, it does not currently provide a simple or inexpensive path to carbon control. In addition to the cost and performance penalties and uncertainties, other factors complicate implementing IGCC carbon control. For example, the nation's largest and least expensive coal supply is western subbituminous coal. However, the IGCC technologies best suited for using this coal also appear to incur the largest cost and performance penalties from adding carbon control technology.[86]

CO_2 Allowance Costs

Congress has considered legislation that would put a cost on carbon emissions, such as the Lieberman-Warner Climate Security Act of 2007 (S. 2191). If Congress ultimately legislates allowance-based carbon controls, the estimated costs of such allowances are very uncertain. As an illustration of this uncertainty, **Figure 5** shows EIA's alternative projections of CO_2 allowance prices under S. 2191. Depending on assumptions for such factors as the speed with which new technologies are deployed and their costs, and the availability for purchase of international CO_2 emission offsets, EIA's estimate of the price of allowances by 2030 ranges from about \$60 to \$160 per metric ton of CO_2 (2006 dollars).

Even the low end of EIA's allowance price forecasts would impose costs far beyond those of existing air emissions regulations. **Figure 6** compares the price of coal in EIA's long-term Reference Case projection (which assumes only current law, and therefore no carbon controls) to EIA's "core" case estimate of allowance prices from the S. 2191 study. Based on EIA's forecasts, by 2030 the allowance price is the equivalent of triple the coal price.[87] (As noted above, the outlook for CO_2 allowance prices is uncertain. Different legislative approaches and changes to other forecasting assumptions can produce very different estimates from those shown here.)

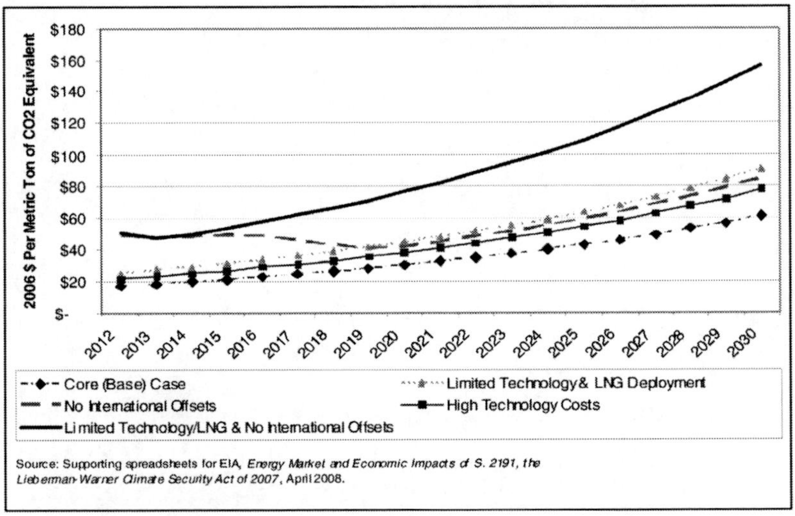

Figure 5. EIA's Projections of S. 2191 CO2 Allowance Prices (2006$ per Metric Ton of CO_2 Equivalent)

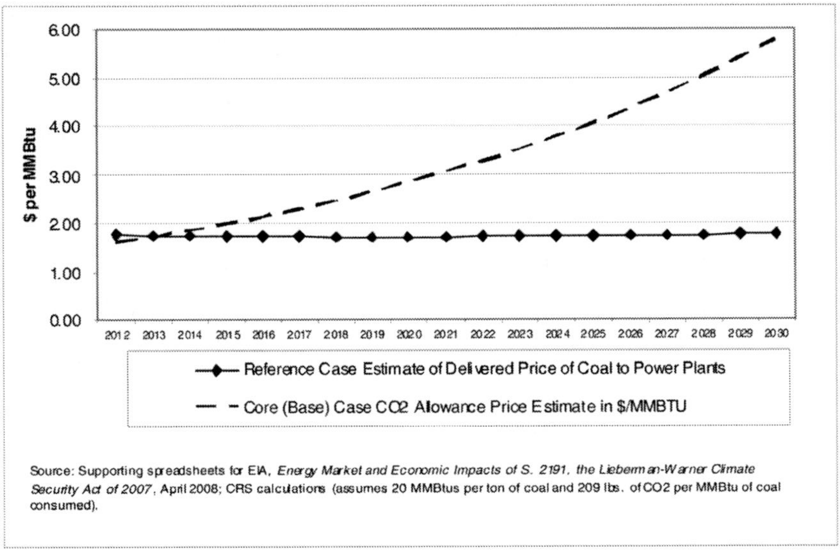

Figure 6. Comparison of EIA's Reference Case Coal Prices and S. 2191 Core Case CO_2 Allowance Prices

Chapter 4

FINANCIAL ANALYSIS METHODOLOGY AND KEY ASSUMPTIONS

This financial analysis of new power plants provides estimates of the operating costs and required capital recovery of each generating technology through 2050. Plant operating costs will vary from year to year depending, for example, on changes in fuel prices and the start or end of government incentive programs. To simplify the comparison of alternatives, these varying yearly expenses are converted to a uniform annualized cost expressed as 2008 present value dollars.

Converting a series of cash flows to a financially equivalent uniform annual payment is a two-step process. First, the cash flows for the project are converted to a 2008 "present value." The present value is the total cost for the analysis period, adjusted ("discounted" using a "discount factor") to account for the time value of money and the risk that projected costs will not occur as expected. This lump-sum 2008 present value is then converted to an equivalent annual payment using a uniform payments factor.[88]

The capital costs for the generating technologies are also converted to annualized payments. An investor-owned utility or independent power producer must recover the cost of its investment and a return on the investment, accounting for income taxes, depreciation rates, and the cost of money. These variables are encapsulated within an annualized capital cost for a project computed using a "capital charge rate." The financial model used for this study computes a project-specific capital charge rate that reflects the assumed cost of money, depreciation schedule, book project life, financing structure (percent debt and percent equity), and composite federal and state income tax rate. For a POU project, which is 100% debt financed, a "capital recovery factor" reflecting each project's cost of money is computed and used to calculate a mortgage-type annual payment.[89]

Combining the annualized capital cost with the annualized operating costs yields the total estimated annualized cost of a project. This annualized cost is divided by the projected yearly output of electricity to produce a cost per Mwh for each technology. By annualizing the costs in this manner, it is possible to compare alternatives with different year-to-year cost patterns on an apples-to-apples basis.

Inputs to the financial model include financing costs, forecasted fuel prices, non-fuel operations and maintenance expense, the efficiency with which fossil-fueled plants convert fuel to electricity, and typical utilization rates (see Appendix D, **Table 17 through Table 20**, below). Most of these inputs are taken from published sources, such as the assumptions EIA used to produce its 2007 and 2008 long-term energy forecasts. The power plant capital costs are estimated by CRS based on a review of public information on recent projects. Appendixes B and C of the report displays the data used for the capital costs estimates.

Chapter 5

ANALYSIS OF POWER PROJECT COSTS

This section of the report analyzes the cost of power from the generating technologies discussed above. Results are first presented for a Base Case analysis. Results are then presented for four additional cases, each of which explores a key variable that influences power plant costs. These cases are:

- Influence of federal and state incentives.
- Higher natural gas price.
- Uncertainty in capital costs.
- Carbon controls and costs.

In each case the cost of power from a natural gas-fired combined cycle plant is used as a benchmark for evaluating the cost of power from the other generating technologies. The gas-fired combined cycle plant is used as a benchmark because of the dominant role it has played, and may continue to play, as the source of new generating capacity capable of meeting baseload and intermediate demand. The closer a generating technology comes to meeting or beating the power cost of the combined cycle, the better its chances of competing in the market for new power plants.

The Base Case is a starting point for comparing how different assumptions, such as for fuel and construction costs, change estimated power costs. None of the cases is a "most likely" estimate of future costs. Future power costs are subject to so many variables with high degrees of uncertainty that projecting a most likely case is impractical. The object of the analysis is provide insight into how key factors influence the costs of power plants, including factors under congressional control such as incentive programs.

These estimates are approximations subject to a high degree of uncertainty. The rankings of the technologies by cost are therefore also an approximation and should not be viewed as definitive estimates of the relative cost-competitiveness of each option. Also note that project-specific factors would weigh into an actual developer's decisions, including how close a fossil plant would be to fuel sources, local climate (for wind and solar), the need for and cost of transmission upgrades, the developer's appetite for risk, and the developer's financial resources.

Case 1: Base Case

Key Obsearvations

- The lowest cost generating technologies in the Base Case are pulverized coal, geothermal, and natural gas combined cycle plants. All have costs around $60 per Mwh (2008 dollars). Based on the assumptions in this report, other technologies are at least a third more expensive.
- Of the three lowest cost technologies, geothermal plants are limited to available sites in the West that typically support only small plants, and coal plants have become harder to build due to cost and environmental issues. The gas-fired combined cycle plant is currently a technology that can be built at a large scale, for cycling or baseload service, throughout the United States.
- The above projections are based on private (IOU or IPP) funding of power projects. The cost per Mwh drops precipitously if the developer is assumed to be a POU with low-cost financing. However, most POUs are small and do not have the financial or managerial resources to build large power projects.

Discussion

As noted earlier in the report, power plants can be built by investor-owned utilities (IOUs), publicly owned utilities (POUs), or independent power producers (IPPs). The Base Case assumes that coal and nuclear plants are constructed by IOUs because they are most likely to have the financial resources and regulatory support to undertake these very large and expensive projects. The natural gas combined cycle plant is assumed to be built by an

IPP. IPPs often prefer to build and operate gas-fired projects because of their relatively low capital costs. The wind, solar, and geothermal plants are also assumed to be IPP projects. The most common current practice is for IPPs to develop renewable projects and sell the power to regulated utilities.

The Base Case has the following characteristics:

- The analysis is for new projects beginning operation in 2015.
- Estimates of fuel prices, allowance prices, and most operational characteristics are from EIA's Reference Case assumptions for the 2008 *Annual Energy Outlook.*[90]
- The 2008 overnight capital costs for each technology are estimated by CRS from public information on recent projects (see **Appendix B**).
- The Base Case excludes "discretionary" incentives: The federal loan guarantee program and clean coal tax credit programs, state utility commission decisions to allow CWIP in rates, and the federal renewable energy production tax credit, which is scheduled to expire at the end of 2010. These incentives are excluded because they are granted by government entities based on a case-by-case analysis of individual projects, and/or are dependent on congressional action to fund or extend the incentives. Accordingly, there is no certainty that most projects will receive these incentives. For example, as of November 2008, DOE had received requests from nuclear plant developers for $122 billion in loan guarantees, compared to congressional approval of only $18.5 billion for nuclear projects.[91]
- The only incentives included in the Base Case are (1) the 30% investment tax credit for solar and geothermal energy systems, which has been extended to 2017 and is automatically available to any qualifying facility; and (2) the nuclear production tax credit, which is available to any qualifying facility. As discussed above, the assumed value of the nuclear credit is 1.35 cents per kWh.
- The Base Case includes no carbon emission controls or costs.

Given these assumptions, **Table 4** presents the resulting annualized cost of power per Mwh for each technology.

Under the Base Case assumptions, the lowest-cost options are pulverized coal, natural gas combined cycle, and geothermal generation, all in the $60 per Mwh (2008 dollars) range (column 10). These results are attributable to the following factors:

- Pulverized coal is a mature technology that relies on a relatively low cost fuel.
- Natural gas is an expensive fuel, but combined cycle technology is highly efficient and has a low construction cost.
- Geothermal energy has no fuel cost and unlike variable renewable technologies, such as wind and solar, can operate at very high utilization rates (high utilization allows the plant to spread fixed operating costs and capital recovery charges over many megawatt- hours of sales).

Although all three technologies have similar power costs, the coal and geothermal technologies have limitations and risks that the natural gas combined cycle does not face. Geothermal plants are limited to relatively small facilities (about 50 MW) at western sites. As discussed above, many coal projects have been canceled due to environmental opposition and escalating construction costs. In contrast, the gas-fired combined cycle plant has limited environmental impacts, can be located wherever a gas pipeline with sufficient capacity is available, and plants can be built with generating capacities in the hundreds of megawatts. Probably the main risk factor for a combined cycle plant is uncertainty over the long term price and supply of natural gas.

In the Base Case, wind power, IGCC coal, and nuclear energy have costs in the $80 per Mwh range. IGCC and nuclear plants are very expensive to build, with estimated overnight capital costs of, respectively, $3,359 and $3,682 per kW of capacity (2008 dollars; see Table 18). Because the plants are expensive and take years to construct (an estimated four years for an IGCC plant and six years for a nuclear plant) these technologies also incur large charges for interest during construction that must be recovered in power costs.

Wind has a relatively high cost per Mwh because wind projects have high capital costs ($2,100 per kW of capacity) and are assumed to operate with a capacity factor of only 34%. The low capacity factor means that the plant is the equivalent of idle two-thirds of the year. Consequently, the capital costs for the plant must be recovered over a relatively small number of units of electricity production, driving up the cost per Mwh. High capital costs and low rates of utilization also drive up the costs of the solar thermal and solar PV plants to, respectively, $100 per Mwh and $255 per Mwh.

Table 4. Estimated Base Case Results (2008 $)

Technology (1)	Developer Type (2)	Non-Fuel O&M Cost (3)	Fuel Cost (4)	SO_2 and NOx Allowance Cost (5)	CO_2 Allow. Cost (6)	Prod.Tax Credit (7)	Total Operating Costs (8)	Capital Reurn (9)	Total Annualized $/Mwh (10)
Coal: Pulverized	IOU	$5.57	$11.13	$0.61	$0.00	$0.00	$17.31	$45.79	$63.10
Coal: IGCC	IOU	$5.46	$10.41	$0.10	$0.00	$0.00	$15.97	$67.02	$82.99
NG: Combined Cycle	IPP	$2.57	$30.57	$0.14	$0.00	$0.00	$33.27	$28.50	$61.77
Nuclear	IOU	$6.13	$5.29	$0.00	$0.00	($3.18)	$8.23	$74.99	$83.22
Wind	IPP	$6.67	$0.00	$0.00	$0.00	$0.00	$6.67	$74.07	$80.74
Geothermal	IPP	$13.69	$0.00	$0.00	$0.00	$0.00	$13.69	$45.54	$59.23
Solar: Thermal	IPP	$13.71	$0.00	$0.00	$0.00	$0.00	$13.71	$86.61	$100.3
Solar: Photovoltaic	IPP	$4.17	$0.00	$0.00	$0.00	$0.00	$4.17	$251.24	$255.4

Source: CRS estimates.

Note: Projections are subject to a high degree of uncertainty. These results should be interpreted as indicative given the projection assumptions rather than as definitive estimates of future outcomes. Mwh = megawatt-hour; IGCC = integrated gasification combined cycle; NG = natural gas; CCS = carbon capture and sequestration; SO_2 = sulfur dioxide; NOx = nitrogen oxides; O&M = operations and maintenance; IPP = independent power producer; IOU = investor owned utility.

Comparison to a Benchmark Price of Electricity

Another way of viewing the results is to compare each technology's costs to a benchmark cost of electricity. As discussed above, the benchmark used is the cost of power from a natural gas combined cycle plant.

Table 5. Benchmark Comparison to Natural Gas Combined Cycle Plant Power Costs: Base Case Values

Technology (1)	Developer Type (2)	Difference in the Power Cost Compared to the Combined Cycle Plant (3)
Geothermal	IPP	-4%
Coal: Pulverized	IOU	2%
Wind	IPP	31%
Coal: IGCC	IOU	34%
Nuclear	IOU	35%
Solar: Thermal	IPP	62%
Solar: Photovoltaic	IPP	313%

Source: CRS estimates.

Note: A negative number indicates that the technology has a power cost lower than that of the combined cycle. Projections are subject to a high degree of uncertainty. These results should be interpreted as indicative given the projection assumptions rather than as definitive estimates of future outcomes. IGCC = integrated gasification combined cycle; IPP = independent power producer; IOU = investor owned utility.

Column 3 of **Table 5** shows the difference between the Base Case power cost for each technology and the Base Case cost of power from the gas-fired combined cycle. Geothermal energy and pulverized coal are the only technologies that have power costs similar to the natural gas combined cycle plant. Nuclear, wind, and coal IGCC power are projected to have costs 31% to 35% higher, and solar thermal has a projected power cost 62% higher. Solar photovoltaic is over 300% higher.

Effect of Financing Costs

The cost of money can have a significant impact on the cost of power. As discussed earlier, POUs have access to lower cost financing than IOUs or IPPs. The significance of lower cost financing is illustrated in **Table 6**, which compares the cost of power assuming IOU and IPP financing (column 3) with the cost of power assuming POU financing (column 4). Excluding for the moment the solar technologies, the reduction in the cost of power ranges from 14% for the combined cycle plant (the least capital-intensive option, which makes it least sensitive to financing costs) to 37% for the capital-intensive IGCC and nuclear plants (column 5). The low cost of public financing helps explain why many capital intensive coal and nuclear projects have POU co-owners.[92]

The reduction in cost by using public financing is only 11% for the solar thermal plant and 14% for the solar photovoltaic plant. The reductions are small because when the plants are publicly financed they lose the 30% renewable energy investment tax credit (POUs do not pay taxes and so cannot take advantage of any tax-based incentives). The loss of the tax credit largely negates the benefit of lower cost POU financing for solar projects.

Case 2: Influence of Federal and State Incentives

Key Observations

- Government financial incentives can make high-cost technologies into low-cost options. The incentive with the greatest impact is the federal loan guarantee, which reduces the cost of financing capital-intensive technologies. With a loan guarantee the cost of nuclear power flips from a high-cost option ($83.22 per Mwh) to one of the low cost ($63.73 per Mwh).
- Even when competing technologies have the advantage of the discretionary government incentives, no technology currently has a significant cost advantage over the natural gas combined cycle.

Discussion

The Base Case includes only non-discretionary incentives: The renewable energy investment tax credit and the nuclear production tax credit. This analysis includes the following discretionary incentives:

- Federal loan guarantees for nuclear power.
- A clean coal tax credit for the IGCC plant.
- A production tax credit for wind (assumes continuation of the terms and conditions of the current production tax credit).
- Return on construction work in progress (CWIP) in rates for IOUs.

Table 6. Effect of Public Power Financing on Base Case Results (2008 $)

Technology (1)	Developer (2)	Annualized Cost per Mwh (3)	Annualized Cost Per Mwh Assuming POU Deeloper (4)	Percent Difference (5)
Coal: Pulverized	IOU	$63.10	$43.97	-30%
Coal: IGCC	IOU	$82.99	$52.44	-37%
NG: Combined	IPP	$61.77	$53.35	-14%
Nuclear	IOU	$83.22	$52.25	-37%
Wind	IPP	$80.74	$54.41	-33%
Geothermal	IPP	$59.23	$47.40	-20%
Solar: Thermal	IPP	$100.32	$89.24	-11%
Solar: Photovoltaic	IPP	$255.41	$219.02	-14%

Source: CRS estimates.

Note: Projections are subject to a high degree of uncertainty. These results should be interpreted as indicative given the projection assumptions rather than as definitive estimates of future outcomes. IGCC = integrated gasification combined cycle; NG = natural gas; Mwh = megawatt-hour; IPP = independent power producer; IOU = investor owned utility; POU = publicly owned utility.

Table 7 shows the effect of the discretionary incentives compared to the Base Case. The additional incentives have the greatest effect on nuclear power. The annualized cost of nuclear generation drops by 23% (column 7), from one of the highest to one of the lowest costs. The most important driver for the nuclear plant is the federal loan guarantee, which allows a developer to fund a project with 80% debt at a much reduced interest rate. The loan guarantee alone cuts the cost of nuclear power by 20% ($15.44 per Mwh).

The renewable production tax credit reduces the cost of wind power by 10%. Geothermal and combined cycle plants (with no additional incentives) and coal (with a 5% reduction in cost due to CWIP in rates) remain low-cost options.

Table 7. Power Costs with Additional Government Incentives (2008 $)

Technology (1)	Developer (2)	Government Incentives in the Base Case (3)	Annualized Cost per Mwh in Base Case (4)	Additional Government Incentives (5)	Annualized Cost Per Mwh With Additional Incentives (6)	Percent Difference (7)
Coal: Pulverized	IOU	None	$63.10	CWIP in rates.	$60.02	-5%
Coal: IGCC	IOU	None	$82.99	ITC; CWIP in rates.	$73.28	-12%
NG: Combined Cycle	IPP	None	$61.77	None	$61.77	0%
Nuclear	IOU	PTC	$83.22	Loan guarantee; CWIP in rates.	$63.73	-23%
Wind	IPP	None	$80.74	PTC	$72.79	-10%
Geothermal	IPP	ITC	$59.23	None	$59.23	0%
Solar: Thermal	IPP	ITC	$100.32	None	$100.32	0%
Solar: Photovoltaic	IPP	ITC	$255.41	None	$255.41	0%

Source: CRS estimates.

Notes: Projections are subject to a high degree of uncertainty. These results should be interpreted as indicative given the projection assumptions rather than as definitive estimates of future outcomes. IGCC = integrated gasification combined cycle; NG = natural gas; Mwh = megawatt-hour; IOU = investor owned utility; IPP = independent power producer; POU = publicly owned utility; PTC = production tax credit; CWIP = construction work in progress; ITC = investment tax credit.

Table 8. Benchmark Comparison to Combined Cycle Power Costs: Additional Government Incentives

Technology (1)	Developer Type (2)	Difference in Power Cost from Combined Cycle	
		Base Case (3)	Additional Incentives (4)
Coal: Pulverized	IOU	2%	-3%
Wind	IPP	31%	18%
Coal: IGCC	IOU	34%	19%
Nuclear	IOU	35%	3%

Source: CRS estimates.

Note: The table only includes the four technologies that receive additional incentives (see Table 7). A negative number indicates that the technology has a power cost lower than that of the combined cycle. Projections are subject to a high degree of uncertainty. These results should be interpreted as indicative given the projection assumptions rather than as definitive estimates of future outcomes. IOU = investor owned utility; IPP = independent power producer.

Table 8 compares the combined cycle benchmark cost of power (column 3) to the cost of power with discretionary incentives (column 4). The table is limited to the technologies that receive the additional incentives: Pulverized coal (CWIP in rates), IGCC coal (CWIP and an investment tax credit), wind (production tax credit), and nuclear (loan guarantee and CWIP). With discretionary incentives, nuclear power swings from a 35% higher cost than the combined cycle to only a 3% difference (comparing columns 3 and 4). The cost advantage of the combined cycle over wind and IGCC coal drops from more than 30% to just under 20%. The cost of power from pulverized coal remains similar to that of the combined cycle.

Case 3: Higher Natural Gas Prices

Key Observations

- If the price of natural gas is assumed to be 50% higher than in the Base Case, geothermal and pulverized coal power are clearly less costly than the combined cycle. However, the use of the geothermal power is limited to available sites in the western United States, and pulverized coal by construction cost and environmental issues.
- In the higher gas price case, the cost of power from the natural gas combined cycle plant converges with wind, nuclear, and IGCC coal. The combined cycle plant no longer has a clear economic advantage over these technologies, but neither is it at a great disadvantage.

Discussion

The economics of natural gas-fired generation pivot on fuel prices. For the base assumptions used in this study, fuel constitutes half of the total cost of power from a new combined cycle power plant, compared to 18% for a coal plant and 6% for a nuclear plant. In addition to being critical to the cost of gas-fired power, natural gas prices are also one of the most uncertain elements in this analysis. As discussed earlier in this report, natural gas prices have been exceptionally difficult to forecast. If the United States becomes more dependent in the future on imports of liquefied natural gas, the domestic and international natural gas markets will be increasingly linked, adding an additional element of uncertainty to the natural gas price outlook.[93]

Underestimates of natural gas prices were pervasive among government and private forecasters in the 1990s and contributed to over-investment in gas-fired generating capacity.[94] If future gas prices are higher than assumed in this report's Base Case, the economics of gas-fired generation could change substantially. The gas market has historically been volatile. Gas prices increased more than 200% from the early 1990s through 2007, and annual increases sometimes exceeded 50% (**Figure 7**).

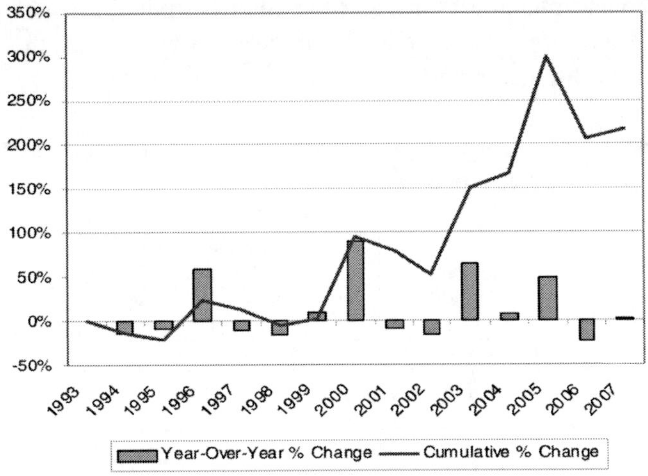

Figure 7. Natural Gas Price Trends (Henry Hub Spot Price)

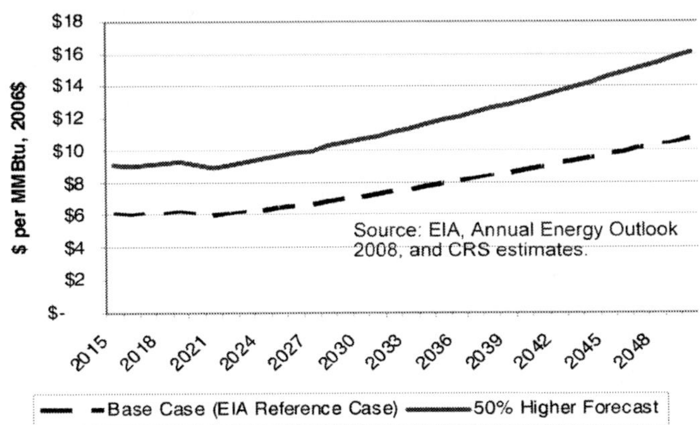

Figure 8. Projection of Natural Gas Prices to Electric Power Plants, 2006 $ per MMBtu

Figure 8 illustrates the Base Case gas price projection and an alternative that ramps up to a level 50% higher than in the Base Case. In the Base Case the annualized cost of power from a natural gas combined cycle plant is $61.77 per Mwh. With a 50% higher gas price, the combined cycle power cost is $77.05 per Mwh. At this power cost the combined cycle is substantially more costly than pulverized coal or geothermal power, and has a clear economic advantage only over the solar technologies (**Table 9**, column 4). On the other hand, even with this much higher fuel price projection, the cost of power from the combined cycle is still comparable to that of wind, nuclear, and IGCC coal generation; and while pulverized coal and geothermal power have lower costs, as discussed above the former is increasingly hard to build for cost and environmental reasons, and the latter is limited to small plants at western sites. Therefore, even with a 50% increase in fuel prices, the gas-fired combined cycle is still a competitive option for new generating capacity.

Table 9. Benchmark Comparison to Natural Gas Combined Cycle Plant Power Costs: 50% Higher Gas Price

Technology (1)	Developer Type (2)	Difference in Power Cost from Combined Cycle Plant	
		Base Case (3)	50% Higher Natural Gas Price (4)
Geothermal	IPP	-4%	-22%
Coal: Pulverized	IOU	2%	-18%
Wind	IPP	31%	5%
Coal: IGCC	IOU	34%	8%
Nuclear	IOU	35%	8%
Solar: Thermal	IPP	104%	30%
Solar: Photovoltaic	IPP	432%	231%

Source: CRS estimates.

Note: A negative number indicates that the technology has a power cost lower than that of the combined cycle. Projections are subject to a high degree of uncertainty. These results should be interpreted as indicative given the projection assumptions rather than as definitive estimates of future outcomes. IGCC = integrated gasification combined cycle; IOU = investor owned utility; IPP = independent power producer.

Another perspective is to determine the increase in the Base Case natural gas price projection required for the cost of power from the natural gas combined cycle plant to equal the cost of power from an alternative technology. This is illustrated in Table 10. The table shows that the price of gas would have to be between 62% to 69% higher than in the Base Case for the cost of power from a combined cycle to equal the projected cost of electricity from nuclear, wind, or coal IGCC technologies (column 3). Natural gas prices would have to increase by about 125% to 635% for the cost of combined cycle power to match solar thermal or solar photovoltaic electricity costs.

Table 10. Change in the Base Case Gas Price Needed to Equalize the Cost of Combined Cycle Power with Other Technologies

Technology (1)	Developer Type (2)	Change in the Base Case Price of Natural Gas Needed to Equalize the Cost of Combined Cycle Power with Other Technologies (3)
Coal: Pulverized	IOU	5%
Coal: IGCC	IOU	69%
Nuclear	IOU	69%
Wind	IPP	62%
Geothermal	IPP	-8%
Solar: Thermal	IPP	125%
Solar: Photovoltaic	IPP	635%

Source: CRS estimates.

Note: Projections are subject to a high degree of uncertainty. These results should be interpreted as indicative given the projection assumptions rather than as definitive estimates of future outcomes. IGCC = integrated gasification combined cycle; IOU = investor owned utility; IPP = independent power producer.

Case 4: Uncertainty in Capital Costs

Key Observations

- Because of its low capital costs and assumed high utilization rate, the power cost of the gas-fired combined cycle plant is about half as sensitive to changes in capital costs as the other technologies.
- The implication is that if power plant capital costs continue to increase rapidly, the competitive position of the combined cycle will improve compared to all other technologies.
- If capital costs decline, the competitive position of the other technologies will substantially improve versus the combined cycle. However, even assuming a 25% drop in capital costs compared to the Base Case, the combined cycle is still competitive with all other technologies.

Discussion

As noted above, the cost of building power plants has recently increased dramatically. Whether costs will continue to increase, remain steady in real dollar terms, or decline is unknown. **Table 11** illustrates the effect on the cost of power of assuming a uniform 25% increase or decrease in capital costs for all technologies compared to the Base Case. Power costs change by about +/- 20% for each technology except for the gas-fired combined cycle plant (+/- 12%; see column 3). This is because the combined cycle has a relatively low capital cost and a high capacity factor.

Table 11. Effect of Higher and Lower Capital Costs on the Cost of Power

Technology (1)	Developer (2)	Change in Cost of Power for a 25% Increase or Decrease in Capital Costs (3)
Coal: Pulverized	IOU	+/-18%
Coal: IGCC	IOU	+/-20%
NG: Combined Cycle	IPP	+/-12%
Nuclear	IOU	+/-23%
Wind	IPP	+/-23%
Geothermal	IPP	+/-19%
Solar: Thermal	IPP	+/-22%
Solar: Photovoltaic	IPP	+/-25%

Source: CRS estimates.

Notes: Projections are subject to a high degree of uncertainty. These results should be interpreted as indicative given the projection assumptions rather than as definitive estimates of future outcomes. IGCC = integrated gasification combined cycle; NG = natural gas; IOU = investor owned utility; IPP = independent power producer.

Table 11 shows that the power cost of the combined cycle is about half as sensitive to changes in capital costs as the other generating technologies. The implication is that continued rapid escalation in the cost of building power plants will favor the economics of combined cycles. This is illustrated by **Table 12**. In the Base Case (Column 3), the power costs of wind, nuclear, and IGCC coal are about a third higher than the combined cycle. In the high capital cost case (Column 4) the difference widens to almost 50%. On the other hand, decreases in capital costs, whether the result of market forces or government incentives, would reduce the cost of power from the other technologies about twice as much as for the combined cycle. This is illustrated by the low capital cost case (Column 5), in which all the non-solar technologies are within 21% or less of the generating cost of the combined cycle.

Case 5: Carbon Controls and Costs

Key Observations

- The estimates of carbon-related allowance costs and control technology costs used in this analysis are subject to an exceptional degree of uncertainty, including whether Congress will actually pass carbon control legislation. The results of this analysis are therefore equally uncertain.
- With the carbon control assumptions used in this analysis, coal-fired generation is expensive, ranging from about $100 to almost $120 per Mwh. The least expensive options include zero-carbon emission technologies: Geothermal ($59.23 per Mwh), nuclear ($83.22) and wind ($80.74).
- The natural gas combined cycle plant without carbon capture is competitive with the other options, even with allowance costs, at $77.21 per Mwh.
- If the cost and efficiency penalties of carbon capture technologies are assumed to drop by 50%, the gas-fired combined cycle plant with capture has an electricity cost comparable to wind and nuclear power. However, a coal plant with capture is still more expensive than wind or nuclear power.

Discussion

Carbon control legislation is under consideration by the Congress, but there has been no agreement on the structure of a control regime or a timetable for implementation. No power plants have been built with full scale carbon capture equipment. The costs of CO_2 allowances and control systems are therefore very uncertain. Actual costs will depend on the content of final legislation (if any), the development of allowance markets in the United States and abroad, and the evolution of control technologies.

Table 12. Benchmark Comparison to Combined Cycle Power Costs: Higher and Lower Capital Costs

Technology (1)	Developer Type (2)	Difference from the Power Cost of the Combined Cycle		
		Base Case (3)	25% Higher Capital Costs (4)	25% Lower Capital Costs (5)
Geothermal	IPP	-4%	3%	-12%
Coal: Pulverized	IOU	2%	8%	-5%
Nuclear	IOU	35%	48%	18%
Wind	IPP	31%	44%	14%
Coal: IGCC	IOU	34%	45%	21%
Solar: Thermal	IPP	62%	77%	44%
Solar: Photovoltaic	IPP	313%	362%	252%

Source: CRS estimates

Note: A negative number indicates that the technology has a power cost lower than that of the combined cycle. Projections are subject to a high degree of uncertainty. These results should be interpreted as indicative given the projection assumptions rather than as definitive estimates of future outcomes. IGCC = integrated gasification combined cycle; IOU = investor owned utility; IPP = independent power producer.

The carbon capture power cost analysis for this study is based on the following assumptions:

- Power plant cost and performance with carbon controls assume current (petrochemical industry based) technology capable of removing 90% of the CO_2. As discussed above, the cost of carbon capture for power plants using petrochemical industry derived technology will be very high. **Table 13** provides estimates of how the capital costs and heat rates of coal and gas plants increase with the addition of carbon controls based on current technology. Capital costs increase by 42% to 97% (column 4), and heat rates increase by 21% to 27% (column 7) resulting in a decline in efficiency. Newer technologies may be less costly and more efficient, but these are still in development.

Table 13. Effect of Current Technology Carbon Controls on Power Plant Capital Cost and Efficiency (2008 $)

Technology (1)	Capital Cost for a Plant Entering Service in 2015 (2008$/kW)			Heat Rate for a Plant Entering Service in 2015 (btus/kWh)		
	Base Case (2)	With Carbon Controls (3)	Percent Change (4)	Base Case (5)	With Carbon Controls (6)	Percent Change (7)
Coal Technologies						
Coal:	$2,48	$3,935	58%	9,11	11,579	27%
Coal: IGCC	$3,35	$4,774	42%	8,52	10,334	21%
Natural Gas Technologies						
NG: Combined Cycle	$1,186	$2,342		6,647	8,332	25%

Source: Table 18.

Note: A higher heat equates to less efficient, and therefore more costly operation. IGCC = integrated gasification combined cycle; NG = natural gas; kW =kilowatt; kWh = kilowatt-hour. Projections are subject to a high degree of uncertainty. These results should be interpreted as indicative given the projection assumptions rather than as definitive estimates of future outcomes.

- The CO_2 allowance price projection is adapted from the EIA "core" case forecast from its analysis of S. 2191.[95] Allowance costs begin in 2012 at $17.70 per metric ton of CO_2 (2008 dollars); increase by 2020 and 2030 to, respectively, $31.34 and $63.99; and reach $266.80 by 2050 (see Table 20 in Appendix D). All allowances must be purchased (i.e., there is no free distribution of allowances to power plants).
- Fuel prices are the same prices used in the Base Case (see **Table 20** in **Appendix D**).
- As in the Base Case, the only financial incentives included are the nuclear production tax credit and the investment tax credit for solar and geothermal plants.

- From a financing standpoint, units with carbon controls are assumed to be high risk projects that incur financing costs equivalent to below investment grade interest rates. This assumption is made because units coming on-line in 2015, as assumed for this study, would be part of the first wave of power plants with carbon controls.

Table 14, below, shows estimates of the levelized cost of power for a carbon capture case.

The results indicate:

- The power costs for coal plants using control technologies are high compared to the Base Case. The costs in the carbon case range from \$100.69 per Mwh to almost \$120 per Mwh (column 10), compared to \$63.19 per Mwh for a pulverized coal unit in the Base Case (**Table 14**, column 10). This illustrates the impact of the high capital costs and efficiency penalties of current carbon capture technologies.
- With the imposition of carbon costs on fossil plants, three of the least expensive options are zero-carbon technologies: Geothermal (\$59.23 per Mwh), nuclear (\$83.22) and wind (\$80.74). Because geothermal plants are limited to specific sites in the western states, nuclear power (a baseload technology) and wind power (a variable renewable resource) are the zero carbon options with relatively low costs and wide latitude for plant sites.
- A fourth relatively low-cost technology is the natural gas combined cycle plant without carbon capture (\$77.21 per Mwh including allowance costs). The relatively low cost is due to the technology's low capital cost, high capacity factor, and relatively low emissions of CO_2 per megawatt-hour of power generated. As shown in Table 14, the natural gas combined cycle plant without carbon capture incurs allowance costs of \$13.06 per Mwh, which is 61% less than the pulverized coal plant cost of \$33.80 per Mwh (column 6). In other words, for every dollar of allowance costs incurred by a coal plant without capture technology, the combined cycle incurs only about 40 cents in costs.[96]
- Solar thermal power (\$100.32 per Mwh) has a lower cost than fossil plants with carbon capture technology, but is still estimated to be about 20% more expensive than nuclear and wind power.

Table 14. Estimated Annualized Cost of Power with Carbon Controls (2008 $)

Technology (1)	Developer Type (2)	Non-Fuel O&M Cost (3)	Fuel Cost (4)	SO_2 and NOx Allownce Cost (5)	CO_2 Allow. Cost (6)	Prod. Tax Credit (7)	Total Operating Costs (8)	Capital Return (9)	Total Annua-lized $/Mwh (10)
Coal Technologies									
Coal: Pulverized	IOU	$5.57	$11.13	$0.61	$33.80	$0.00	$51.11	$49.58	**$100.69**
Coal: Pulverized/CCS	IOU	$13.48	$14.13	$0.77	$4.29	$0.00	$32.67	$78.87	**$111.54**
Coal: IGCC	IOU	$5.46	$10.41	$0.10	$31.61	$0.00	$47.58	$67.02	**$114.60**
Coal: IGCC/CCS	IOU	$7.10	$12.61	$0.13	$3.83	$0.00	$23.67	$95.25	**$118.92**
Natural Gas Technologies									
NG: Combined Cycle	IPP	$2.57	$30.57	$0.14	$13.06	$0.00	$46.34	$30.88	**$77.21**
NG: Combined Cycle/CCS	IOU	$3.68	$38.32	$0.17	$1.64	$0.00	$43.81	$51.09	**$94.90**
Zero Carbon Technologies									
Geothermal	IPP	$13.69	$0.00	$0.00	$0.00	$0.00	$13.69	$45.54	**$59.23**
Nuclear	IOU	$6.13	$5.29	$0.00	$0.00	($3.18)	$8.23	$74.99	**$83.22**
Wind	IPP	$6.67	$0.00	$0.00	$0.00	$0.00	$6.67	$74.07	**$80.74**
Solar: Thermal	IPP	$13.71	$0.00	$0.00	$0.00	$0.00	$13.71	$86.61	**$100.32**
Solar: Photovoltaic	IPP	$4.17	$0.00	$0.00	$0.00	$0.00	$4.17	$251.24	**$255.41**

Source: CRS estimates.

Note: Projections are subject to a high degree of uncertainty. These results should be interpreted as indicative given the projection assumptions rather than as definitive estimates of future outcomes. Mwh = megawatt-hour; IGCC = integrated gasification combined cycle; NG = natural gas; CCS = carbon capture and sequestration; SO_2 = sulfur dioxide; NOx = nitrogen oxides; O&M = operations and maintenance; IOU = investor owned utility; IPP = independent power producer.

The relatively low cost of power from the natural gas combined cycle plant is in part a function of the fuel price. As noted above, the carbon capture analysis uses the same fuel price projections as in the Base Case. It is possible that in a carbon- constrained world demand for gas will increase, driving up prices. As shown below in **Table 15**:

- A 12% increase in the price of gas would equalize the cost of electricity from the combined cycle plant without carbon capture with wind power (column 3);
- A 20% increase would equalize the power cost of the combined cycle plant and the nuclear plant;
- The price of natural gas would have to more than double for the power cost of the gas-fired combined cycle plant to equal the cost of coal power with carbon controls, or increase by 75% to match the cost of solar thermal power.

Table 15. Change in the Price of Natural Gas Required to Equalize the Cost of Combined Cycle Generation (Without Carbon Controls) with Other Technologies

Technology (1)	Developer (2)	Change in Price of Natural Gas from Base Case Necessary to Equalize Cost of Power (3)
Coal: Pulverized	IOU	77%
Coal: IGCC	IOU	123%
Coal: Pulverized/CCS	IOU	112%
Coal: IGCC/CCS	IOU	136%
Nuclear	IOU	20%
Wind	IPP	12%
Geothermal	IPP	-59%
Solar: Thermal	IPP	75%
Solar: Photovoltaic	IPP	580%

Source: CRS estimates.

Note: Projections are subject to a high degree of uncertainty. These results should be interpreted as indicative given the projection assumptions rather than as definitive estimates of future outcomes. IGCC = integrated gasification combined cycle; NG = natural gas; CCS = carbon capture and sequestration; IOU = investor owned utility; IPP = independent power producer.

Table 16. Cost of Power with Base and Reduced Carbon Capture Cost and Efficiency Impacts

Technology (1)	Carbon Control Base Case				Lower Cost Carbon Controls (50% Lower Capital Costs and Heat Rates)			
		% Difference from:				% Difference from:		
	Power Cost (2008 $/ Mwh) (2)	Cost of Gas Fired Comb-ined Cycle *without* CCS (3)	Cost of Nuclear Power (4)	Cost of Wind Power (5)	Power Cost (2008 $/Mwh) (6)	Cost of Gas-Fired Combined Cycle *without* CCS (7)	Cost of Nuclear Power (8)	Cost of Wind Power (9)
Coal Technologies								
Coal: Pulverized/ CCS	$111.54	44%	34%	38%	$96.64	25%	16%	20%
Coal: IGCC/CCS	$118.92	54%	43%	47%	$103.08	34%	24%	28%
Natural Gas Technologies								
NG: Combined Cycle/CCS	$94.90	23%	14%	18%	$77.81	1%	-7%	-4%

Source: CRS estimates.

Note: The estimated costs of combined cycle power without carbon capture, nuclear power, and wind power are, respectively, $77.21, $83.22, and $80.74 per Mwh (2008 dollars). Mwh = megawatt-hour; IGCC = integrated gasification combined cycle; NG = natural gas; CCS = carbon capture and sequestration. Projections are subject to a high degree of uncertainty. These results should be interpreted as indicative given the projection assumptions rather than as definitive estimates of future outcomes.

This scale of natural gas price increases has precedent. As shown in **Figure 7**, between the early 1990s and 2007 the market price of natural gas increased by about 200%.

As discussed above, the cost and efficiency impacts of current carbon capture technologies are high, and improved technologies are under development. **Table 16** shows the estimated cost of power for plants with carbon capture assuming that capital cost and heat rate (efficiency) penalties are both reduced by 50%. In this case the combined cycle plant with capture has an electricity cost slightly less than wind and nuclear power, and the pulverized coal plant with capture closes to within 20% of wind power and 16% of nuclear (columns 8 and 9). The IGCC plant with capture is more expensive, with a power cost 28% higher than wind and 24% higher than nuclear; this result reflects the high cost of IGCC technology even before carbon capture is added.

Appendix A

POWER GENERATION TECHNOLOGY PROCESS DIAGRAMS AND IMAGES

Pulverized Coal

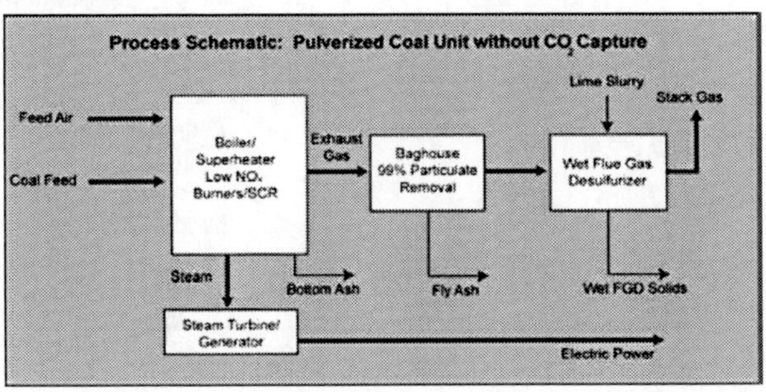

Figure 9. Process Schematic: Pulverized Coal without Carbon Capture

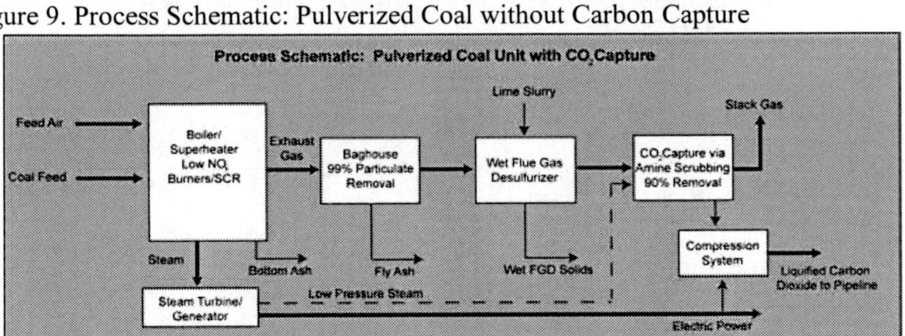

Figure 10. Process Schematic: Pulverized Coal with Carbon Capture

Sources: Image courtesy of Industcards.com; diagrams adapted from MIT, *The Future of Coal,* 2007.

Figure 11. Representative Pulverized Coal Plant: Gavin Plant (Ohio)

Integrated Gasification Combined Cycle Coal (IGCC)

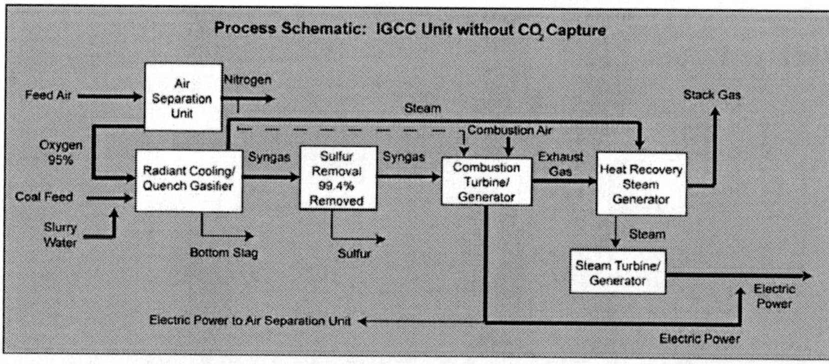

Figure 12. Process Schematic: IGCC without Carbon Capture

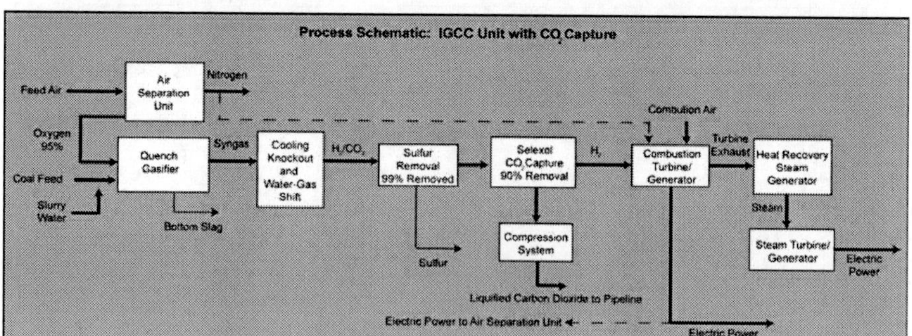

Figure 13. Process Schematic: IGCC with Carbon Capture

Sources: image courtesy of Industcards.com; diagrams adapted from MIT, *The Future of Coal,* 2007.

Figure 14. Representative IGCC Plant: Polk Plant (Florida)

Natural Gas Combined Cycle

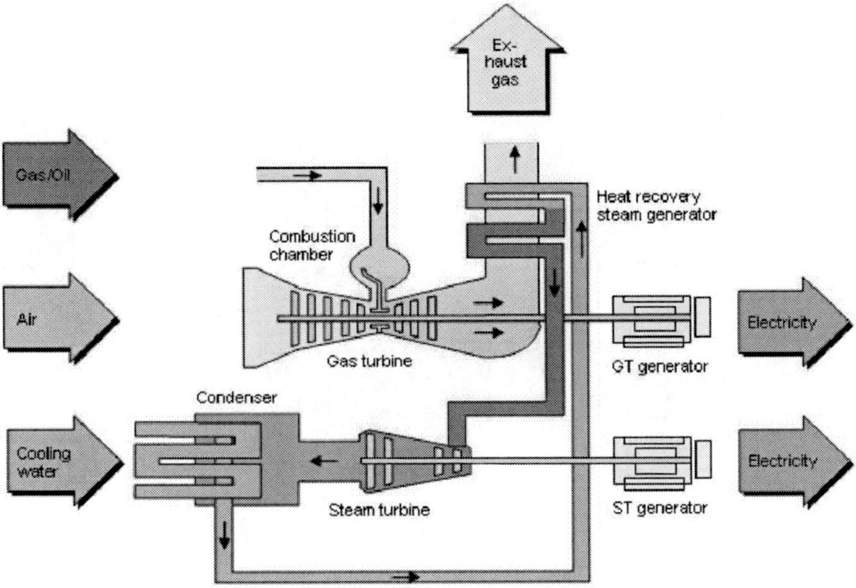

Figure 15. Process Schematic: Combined Cycle Power Plant

Sources: Diagram from Siemens Energy [http://www.powergeneration.siemens.com/products-solutions-services/power-plant-soln/combined-cycle-power-plants/CCPP.htm]; image courtesy of Industcards.com.

Figure 16. Representative Combined Cycle: McClain Plant (Oklahoma)

Nuclear Power

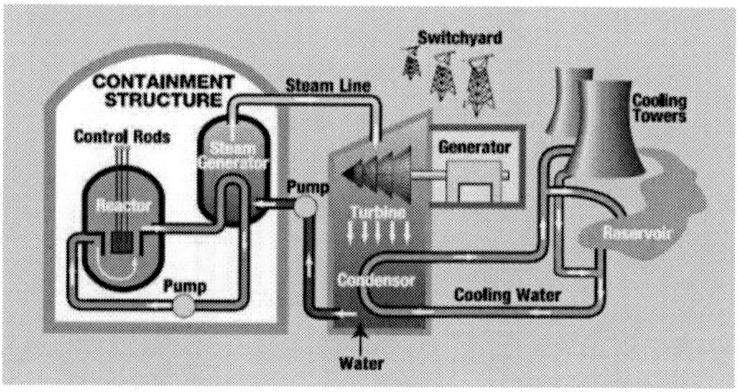

Figure 17. Process Schematic: Pressurized Water Reactor (PWR)

- Water is heated by the fuel rods; the water is kept under high pressure and does not boil.
- The hot water from the reactor passes through tubes inside a steam generator, where the heat is transferred to water flowing around the tubes.
- The water in this secondary loop boils and turns to steam.

- The steam turns the turbines that spin the generator to produce electricity.
- After its energy is used up in the turbines, the steam is drawn into a condenser, where it is cooled back into water and reused.

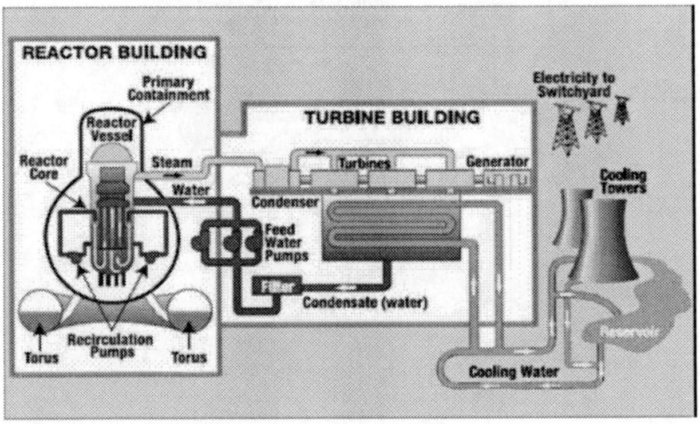

Figure 18. Process Schematic: Boiling Water Reactor (BWR)

- Water is pumped through the reactor and is heated by the fuel rods.
- The water boils, turning to steam.
- The force of the expanding steam drives the turbines, which spin the generator to produce electricity.
- After its energy is used up in the turbines, the steam is drawn into a condenser, where it is cooled back into water and reused.

Sources: Diagrams and accompanying text from Tennessee Valley Authority ([http://www.tva.gov/power/pdf/nuclear.pdf]); AP1000 image from Progress Energy ([http://www.progress-energy.com/aboutenergy/poweringthefuture_florida/levy/ap1000.jpg]).

Figure 19. Representative Gen III/III+ Nuclear Plant: Rendering of the Westinghouse AP1000 (Levy County Project, Florida)

Wind

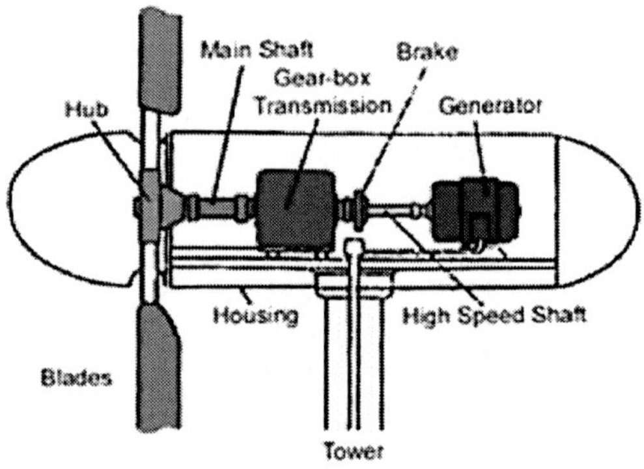

Figure 20. Schematic of a Wind Turbine

Figure 21. Representative Wind Farm: Gray County Wind Farm (Kansas)

Sources: Schematic from California Energy Commission EnergyQuest website (www.energyquest.ca.gov/story/chapter16.html); image of Gray County wind farm from [http://www.kansastravel.org/graycountywindfarm.htm]; image of wind turbine scale from FPL Energy ([http://www.fplenergy.com/renewable/pdf /NatLeaderWind.pdf])

Figure 22. Wind Turbine Size and Scale (FPL Energy)

Geothermal

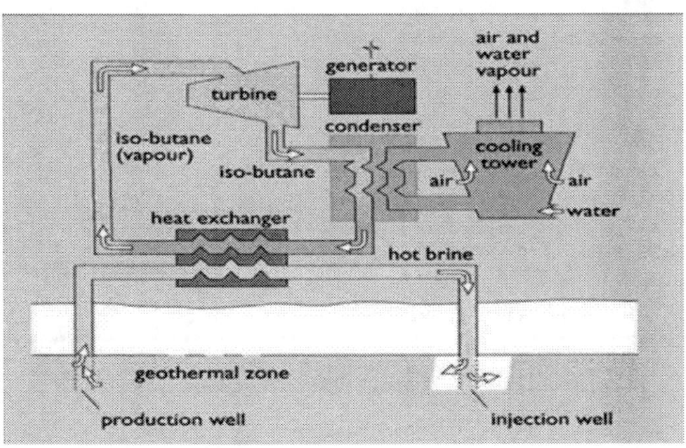

Figure 23. Process Schematic: Binary Cycle Geothermal Plant

Sources: diagram from Steven Lawrence, presentation on "Geothermal Energy," University of Colorado, undated, citing Godfrey Boyle, Renewable Energy, 2nd Edition, 2004 [http://leeds-faculty.colorado.edu/lawrence/syst6820/Lectures/Geothermal%20Energy.ppt]; image courtesy of Industcards.com.

Figure 24. Representative Geothermal Plant: Raft River Plant (Idaho)

Solar Thermal Power

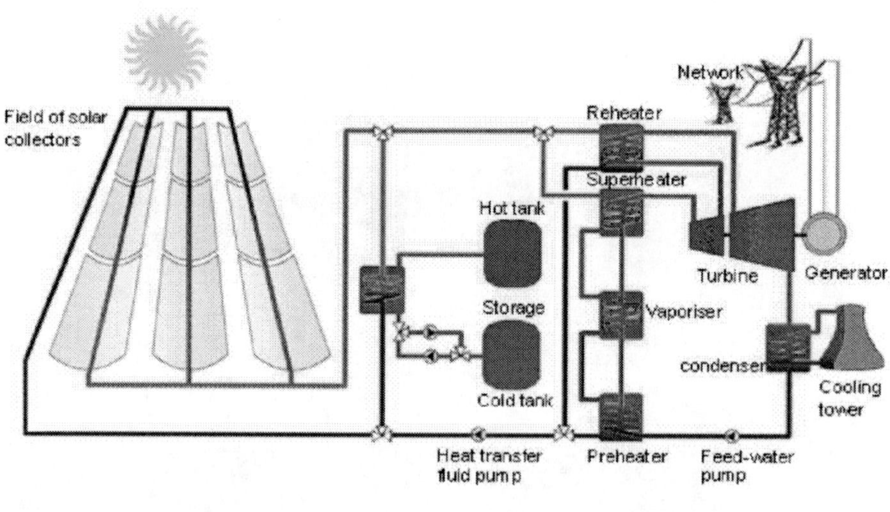

Figure 25. Process Schematic: Parabolic Trough Solar Thermal Plant

Sources: Diagram from [http://www.solarserver.de/solarmagazin/solar-report_0207_e.html]; images from [http://www.solargenixchicago.com/nevadaone.cfm].

Figure 26. Representative Solar Thermal Plant: Nevada Solar One

Sources: Diagram from [http://www.solarserver.de/solarmagazin/solar-report_0207_e.html]; images from [http://www.solargenixchicago.com/nevadaone.cfm].

Figure 27. Nevada Solar One: Parabolic Collector Detail

Solar Photovoltaic Power

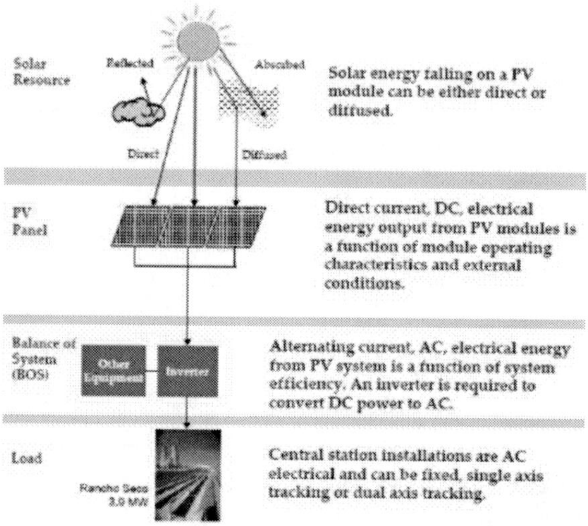

Figure 28. Process Schematic: Central Station Solar Photovoltaic Power

Figure 29. Representative Solar PV Plant: Nellis Air Force Base (Nevada)

Sources: Diagram from California Energy Commission, Comparative Costs of California Central Station Electricity Generation Technologies, Appendix B, p. 61; images from the Nellis Air Force Base website at [http://www.nellis.af.mil/shared/media/document/AFD-080117-039.pdf].

Figure 30. Nellis AFB Photovoltaic Array Detail

Appendix B

ESTIMATES OF POWER PLANT OVERNIGHT COSTS

The financial analysis model used in this study calculates the capital component of power prices based on the "overnight" cost of a power plant. The overnight cost is the cost that would be incurred if a power plant could be built instantly. The overnight cost therefore excludes escalation in equipment, labor, and commodity prices that could occur during the time a plant is under construction. It also excludes the financing charges, often referred to as interest during construction (IDC), incurred while the plant is being built.

With the exception of plants using carbon control technology (see Appendix C) the overnight costs were estimated for this study from public information on actual power projects. The costs were estimated as follows:

- CRS developed a database of information on 161 power projects and cost estimates covering the fossil, nuclear, and renewable energy technologies included in this report.
- A subset of the projects in the database were used to estimate overnight costs. Projects were excluded for many reasons, including because the projects were too old to reflect current construction costs, did not use standard technology, were extreme high or low outliers and no information was available to explain the costs, or had other unusual characteristics (e.g., some plants reduced costs by purchasing used or surplus equipment).

- The remaining projects were sorted by technology (e.g., nuclear, wind, etc.). The reported cost per kilowatt of capacity for the projects in each group were then averaged to estimate the overnight cost for each technology.

To the extent possible the information for the database was taken from information filed by utilities with state public service commissions. The advantage of using this source is that utilities seeking permission to construct new plants are often required to disgorge cost details. With these details the project cost estimate can be adjusted to exclude IDC and other expenses not directly associated with the cost of the plant, such as major transmission system upgrades distant from the plant site.

When utility commission filings for a project were not available, as was almost always true for IPP and POU projects, other public sources were used, including press releases and trade journal articles. In most cases it was possible to determine whether or not a cost estimate included IDC. However, it was rarely possible, with or without utility commission filings, to determine how much cost escalation was built into a project estimate. Because it was not possible to extract the escalation costs from the project estimates, as a rough correction the financial model assumed no cost escalation to avoid a double count. The model does compute the IDC charges.

The 161 projects in the database includes information on 119 United States power plant projects. Some are still in the planning stage, and a few never progressed beyond paper studies and were canceled. The database also includes information on 31 generic and 11 foreign cost estimates for nuclear power plants. (A generic estimate is a cost estimate not associated with any real project or specific site. Generic estimates are usually made by vendors or found in government and academic studies.) The generic and foreign estimates are useful for illustrating cost trends because no nuclear plants have been built in the United States in many years, but none were used in the final estimate of the overnight nuclear plant cost.

Although the capital costs used in this study are based on these actual project estimates, the capital costs are still subject to significant uncertainty due to such as factors as cost escalation and evolution in power plant and construction technology. The uncertainty is greatest for the technologies which have the least commercial experience, such as advanced nuclear plants and IGCC coal plants.

Immediately following is information on the projects used to estimate overnight costs for this report. There is a table for each technology (e.g., pulverized coal) listing each project used to estimate the overnight cost for that technology. Accompanying each table is a graph showing the time trend for that technology's capital costs. The data points on the graph are marked to indicate whether a point represents a project used in estimating the overnight cost, or another project that was excluded from the estimate for one of the reasons discussed above. The time axis for these graphs is the actual or planned first year of commercial service.

The following acronyms are used in the tables:

ABWR:	Advanced boiling water [nuclear] reactor
AP1000:	Advanced Passive 1000 [nuclear reactor]
COD	Commercial Operating Date
ESBWR:	Economic simplified boiling water [nuclear] reactor
IGCC:	Integrated gasification combined cycle [coal]
PT:	Parabolic trough [solar]
PV:	Photovoltaic [solar]
SCPC:	Supercritical pulverized coal
U.S. - EPR:	United States - Evolutionary Pressurized [nuclear] Reactor
UNK:	Unknown
USCPC:	Ultra-supercritical pulverized coal

Pulverized Coal

Pulverized Coal Projects Selected for Cost Estimate (Average Cost per Kw: $2,519; Rounded Average: $2,500)

Plant Name	State	Lead Developer	Type of Ownership	Energy Source	Technology	Net Summer Capacity (Mw)	Cost (million $)	Cost per Kw	COD Year	Greenfield (G) or Brownfield (B)	Sources
Sutherland Generating Station Unit 4	IA	Alliant Energy	Utility	COAL	SCPC	649	$1,854	$2,857	2013	B	Ryberg Williams, "Three Iowa Co-Ops, Wisconsin's Alliant to Own Coal Plant," Des Moines Register, November 29, 2007; Alliant Energy Press Releases, December 10, 2007 and March 312, 2008; Dave DeWitte, "Marshalltown Plant Could Burn Switchgrass," The (Cedar Rapids) Gazette, April 10, 2007.
Pee Dee	SC	South Carolina Public Service Authority (Santee Cooper)	Utility	COAL	SCPC	600	$1,250	$2,083	2012	G	Santee Cooper Press Release, May 22, 2006; Santee Cooper, Draft Environmental Assessment: Pee Dee Electrical Generating Station, October 31, 2006; Tony Bartelme, "Santee Cooper Ups Cost of Coal Plant," The (Charleston) Post and Courier, March 27, 2008.

(Continued)

Plant Name	State	Lead Developer	Type of Ownership	Energy Source	Technology	Net Summer Capacity (Mw)	Cost (million $)	Cost per Kw	COD Year	Greenfield (G) or Brownfield (B)	Sources
Big Stone 2	SD	Otter Tail Power Co.	Utility	COAL	SCPC	580	$1,411	$2,433	2013	B	Supplemental Prefiled Testimony of Mark Rolfes on behalf of Otter Tail Power Co., before the Minnesota Public Utilities Commission, Dockets CN-05-619 and TR05-1275, November 13, 2007.
John W. Turk, Jr. (Hempstead)	AR	Southwestern Electric Power Co.	Utility	COAL	USCPC	609	$1,522	$2,499	2013	G	Texas Public Utilities Commission, Proposal for Decision, Docket 33891, January 17, 2008; Direct Testimonies of Renee Hawkins and James Kobyra on behalf of Southwestern Electric Power Co., before the Texas Public Utilities Commission, Docket 33891, February 20, 2007; Supplemental Direct Testimonies of Renee Hawkins and James Kobyra on behalf of Southwestern Electric Power Co., before the Texas Public Utilities Commission, Docket 33891, April 22, 2008; Housley Carr, "Texas Commission

(Continued)

Plant Name	State	Lead Developer	Type of Ownership	Energy Source	Technology	Net Summer Capacity (Mw)	Cost (million $)	Cost per Kw	COD Year	Greenfield (G) or Brownfield (B)	Sources
											Delays Approval of SWEPCO's 600-MW, Coal- Fired Plant," Platts Electric Utility Week, June 9, 2008.
Cliffside Unit 6	NC	Duke Energy	Utility	COAL	SCPC	800	$1,800	$2,250	2012	B	Law Office of Robert W. Kaylor, on be- half of Duke Energy Carolinas, letters to the North Carolina Utilities Commission, Cliffside Cost Estimates, May 30, 2007 and December 28, 2007; North Carolina Utilities Commission, Decision, Docket E7, Sub 790, March 21, 2007; Duke Energy 10-Q for 3rd quarter 2007, p. 33.
American Mu- nicipal Power Generatin g Station 1 & 2	OH	American Municipal Power - Ohio	Utility	COAL	SCPC	960	$2,950	$3,073	2013	G	R.W. Beck, Initial Project Feasibility Study Update, January 2008 (redacted public version); Direct testimonies of Ivan Clark and Scott Kiesewetter on behalf of American Municipal Power - Ohio, before the Ohio Power Siting Board, Case 06-1358-EL-BGN; American

(Continued)

Plant Name	State	Lead Developer	Type of Ownership	Energy Source	Technology	Net Summer Capacity (Mw)	Cost (million $)	Cost per Kw	COD Year	Greenfield (G) or Brownfield (B)	Sources
											Municipal Power - Ohio, Application to the Ohio Power Siting Board, Case 06-1358-EL-BGN, May 4, 2007.
Holcomb Sta- tion Units 3 and 4	KA	Sunflower Electric Power Corp.	Utility	COAL	SCPC	1,400	$3,600	$2,571	201 2	B	John Hanna, "Supporters Hunt for Votes on Coal Plants as Deadline Looms," Associated Press, 2/20/2008; [http ://www.holcombstation. coop/].
Sandy Creek Energy Station	TX	LS Power	Mixed	COAL	SCPC	900	$2,196	$2,440	201 2	G	"Dynegy, LS Power Ready to Start Con struction of Sandy Creek," Platts Commodity News, 9/4/2007; "Moody's Assigns Ba3 Rating to Sandy Creek Facilities," Moody's Investors Service Press Release, 8/14/2007; Steve Hooks, "LCRA Grabs 22% Stake in Texas Coal Project," Platts Coal Trader, June 11, 2008.

(Continued)

Plant Name	State	Lead Developer	Type of Ownership	Energy Source	Technology	Net Summer Capacity (Mw)	Cost (million $)	Cost per Kw	COD Year	Greenfield (G) or Brownfield (B)	Sources
Norborne	MO	Associated Electric Coopera-tive Inc	Utility	COAL	SCPC	689	$1,700	$2,467	2012	G	Associated Electric Cooperative Press Re lease, 3/3/2008; Missouri Air Conservation Commission, Permit to Construct No. 022008-010, February 22, 2008; Karen Dillon, "Construction of Coal-Fired Power Plant East of Excelsior Springs Delayed Indefinitely," The Kansas City Star, 3/3/08; "Co-op Drops Approved Missouri Coal-Fired Plant Over Unease About CO_2 Rules, Cost," Platts Coal Trader,

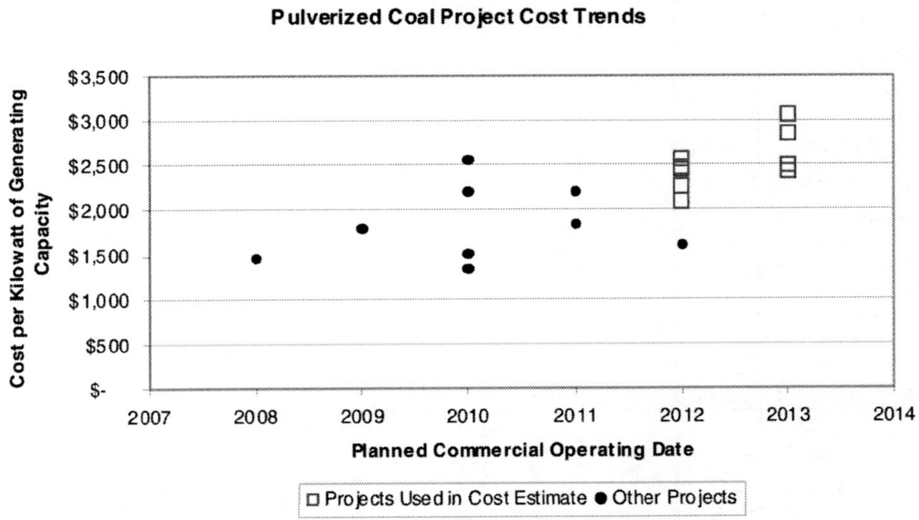
Pulverized Coal Project Cost Trends
Cost per Kilowatt of Generating Capacity
$3,500
$3,000
$2,500
$2,000
$1,500
$1,000
$500
$-
2007
2008
2009
2010
2011
2012
2013
2014
Planned Commercial Operating Date
Projects Used in Cost Estimate
Other Projects

Integrated Gasification Combined Cycle (IGCC) Coal

Coal Integrated Gasification Combined Cycle (IGCC) Projects Selected for Cost Estimate (Average Cost per Kw: \$3,390; Rounded Average: \$3,400)

Plant Name	State	Lead Developer	Type of Ownership	Energy Source	Technology	Net Summer Capacity (Mw)	Cost (million \$)	Cost per Kw	COD Year	Greenfield (G) or Brownfield (B)	Sources
Mountaineer IGCC	WV	American Electric Power	Utility	COAL	IGCC	629	\$2,230	\$3,545	2013	B	"Appalachian Power Says it Would Consider Cap on Construction Costs for IGCC Project," Platts Global Power Report, December 13, 2007; AER Press Release, June 18, 2007; West Virginia Public Service Commission, Case 06-0033-E-CN: Direct testimonies on behalf of Applachian Power Co. of Dana E. Waldo, William M. Jasper, and Terry Eads, June 18, 2007; Final Order, March 6, 2008. "W.VA. Clears AEP's IGCC Project; Commission May Want Cost Justification," Platts Coal Trader, March 10, 2008.

(Continued)

Plant Name	State	Lead Developer	Type of Ownership	Energy Source	Technology	Net Summer Capacity (Mw)	Cost (million $)	Cost per Kw	COD Year	Greenfield (G) or Brownfield (B)	Sources
Great Bend	OH	American Electric Power	Utility	COAL	IGCC	629	$2,200	$3,498	2015	G	Bob Matyi, "Ohio Consumer Advocate Takes Aim at Financing for AEP's Planned IGCC Project," Platts Electric Utility Week, October 15, 2007; Ohio Public Utilities Commission, Opinion and Order, Case 05- 376-EL-UNC, April 10, 2006.
Taylorville Energy Cen ter	IL	Tenaska	IPP	COAL	IGCC	630	$2,000	$3,175	2012	G	"EPA Rejects Challenge to $2B Energy Plant in Central Illinois," Associated Press, January 31, 2008; "Taylorville Energy Center — Facts" [http://www.tenaska.com/userfiles/File/Taylorville%20Fact %20Sheet (1).pdf].
emper County	MS	Southern Company	Utility	COAL	IGCC	600	$1,800	$3,000	2013	G	"Mississippi Power Moving Forward with Plans for Coal Gasification Facillity," U.S. Coal Review, December 18, 2006.

(Continued)

Plant Name	State	Lead Developer	Type of Ownership	Energy Source	Technology	Net Summer Capacity (Mw)	Cost (million $)	Cost per Kw	COD Year	Greenfield (G) or Brownfield (B)	Sources
Edwardsport IGCC	IN	Duke Energy	Utility	COAL	IGCC	630	$2,350	$3,730	2011	B	Indiana Utility Regulatory Commission, Order, Causes 43114 and 43114-S, November 20, 2007; Rebuttal Testimony of Stephen M. Farmer Before the Indiana Utility Regulatory Commission, Causes 43114 and 43114-S, May 31, 2007; Virginia State Corporation Commission, Final Order, Case PUE-2007-00068; Duke Energy press release, May 1, 2008.

IGCC Project Cost Trends

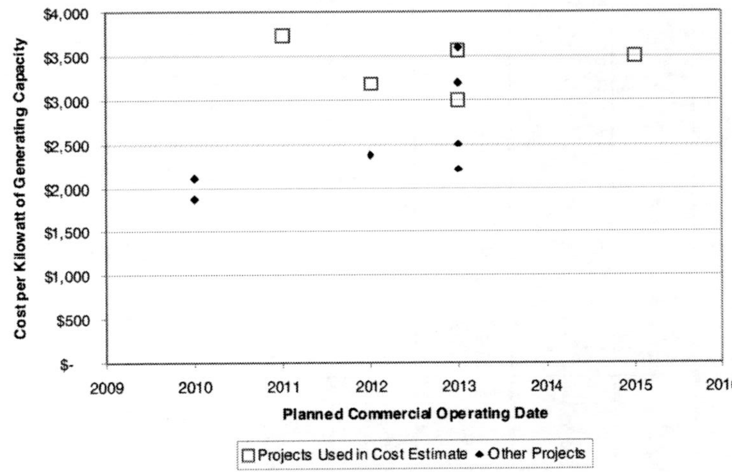

Nuclear

Nuclear Projects Selected for Cost Estimate
(Average Cost per Kw: $3,930; Rounded Average: $3,900)

Plant Name	State	Lead Developer	Type of Ownership	Energy Source	Technology	Net Summer Capaci ty (Mw)	Cost (million $)	Cost per Kw	COD Year	Greenfield (G) or Brownfield (B)	Sources
Calvert Cliffs 3	MD	Constellation	Utility	Nuclear	US-EPR	1,600	$9,194	$5,746	2015	B	Q4 2007 Constellation Energy Group, Inc. Earnings Conference Call, January 30, 2008 — Final (FD Wire); Jeff Beattie, "Constellation Promotes Wallace, Hires Barron to Lead Nuke Charge," The Energy Daily, March 5, 2008; Constellation Energy 2Q 2008 earnings presentation; Application of Unistar Nuclear to the Maryland Public Service Commission for a CCN, 11/13/2007, Case No. 9127.

(Continued)

Plant Name	State	Lead Developer	Type of Ownership	Energy Source	Technology	Net Summer Capacity (Mw)	Cost (million $)	Cost per Kw	COD Year	Greenfield (G) or Brownfield (B)	Sources
Levy County 1&2	FL	Progress Energy Florida	Utility	Nuclear	AP1 000	2,184	$9,304	$4,260	2016	G	Florida PSC Docket 080148-EI: Petition filed by Progress Energy Florida (PEF): Testimonies on behalf of PEF by Daniel L. Roderick (redacted); Javier Portuondo, and John Crisp (including attached Need Determination Study).
South Texas Project Units 3 and 4 - High Estimate	TX	NRG	Utility	Nuclear	ABWR	2,700	$9,909	$3,670	2015	B	"Nuclear Power—- Leading the US Re- vival," Modern Power Systems, 12/13/2007; NRG Press Release, 9/24/2007; NRG Analyst Presentation, "NRG and Toshiba: EmPowering Nuclear Development in US," March 26, 2008; Transcript and audio recording of NRG analyst presentation on formation of Nuclear Innovation North America, March 26, 2008 (transcript from Fair Disclosure Wire, audio recording from NRG website).

(Continued)

Plant Name	State	Lead Developer	Type of ownership	Energy Source	Technology	Net Summer Ca- paci ty (Mw)	Cost (million $)	Cost per Kw	COD Year	Greenfield (G) or Brownfield (B)	Sources
South Texas Project Units 3 and 4 - Low Esti- mate	TX	NRG	Utility	Nuclear	ABWR	2,700	$7,736	$2,865	2015	B	"Nuclear Power — Leading the US Re- vival," Modern Power Systems, 12/13/2007; NRG Press Release, 9/24/2007; NRG Analyst Presentation, "NRG and Toshiba: EmPow- ering Nuclear Development in US," March 26, 2008; Transc- ript and audio recording of NRG analyst presentation on formation of Nuclear Innovat- ion North America, March 26, 2008 (transcript from Fair Disclosure Wire, audio recording from NRG website).
South Texas Project Units 3 and 4 - Middle Esti- mate	TX	NRG	Utility	Nuclear	ABWR	2,700	$8,640	$3,200	2015	B	"Nuclear Power — Leading the US Re- vival," Modern Power Systems, 12/13/2007; NRG Press Release, 9/24/2007; NRG Analyst Presentation, "NRG and Toshiba: EmPowering Nuclear Development in US," March 26, 2008; Transcript and audio recording of NRG analyst

(Continued)

Plant Name	State	Lead Developer	Type of Ownership	Energy Source	Technology	Net Summer Capaci ty (Mw)	Cost (million $)	Cost per Kw	COD Year	Greenfield (G) or Brownfield (B)	Sources
											presentation on formation of Nuclear Innovation North America, March 26, 2008 (transcript from Fair Disclosure Wire, audio recording from NRG website).
Turkey Point 6 & 7 - Case A	FL	Florida Power & Light	Utility	Nuclear	ESBWR or AP-1000	2,200	$7,911	$3,596	2018	B	Direct Testimony of Steven Scroggs on behalf of Florida Power & Light, Florida Public Service Commission Docket 070650-EI, October 16, 2007.
Turkey Point 6 & 7 - Case B	FL	Florida Power & Light	Utility	Nuclear	ESBWR or AP-1000	2,200	$6,838	$3,108	2018	B	Direct Testimony of Steven Scroggs on behalf of Florida Power & Light, Florida Public Service Commission Docket 070650-EI, October 16, 2007.
Turkey Point 6 & 7 - Case C	FL	Florida Power & Light	Utility	Nuclear	ESBWR or AP-1000	2,200	$9,988	$4,540	2018	B	Direct Testimony of Steven Scroggs on behalf of Florida Power & Light and Need Study for Electrical Power, Florida Public Service Commission Docket 070650-EI, October 16, 2007.

(Continued)

Plant Name	State	Lead Developer	Type of Ownership	Energy Source	Technology	Net Summer Capacity (Mw)	Cost (million $)	Cost per Kw	COD Year	Greenfield (G) or Brownfield (B)	Sources
V.C. Summer 2 & 3	SC	South Carolina Electric & Gas	Utility	Nuclear	AP1000	2,234	$9,800	$4,387	2016	B	Joint press release by SCANA Corp. and Santee Cooper, May 27, 2008.

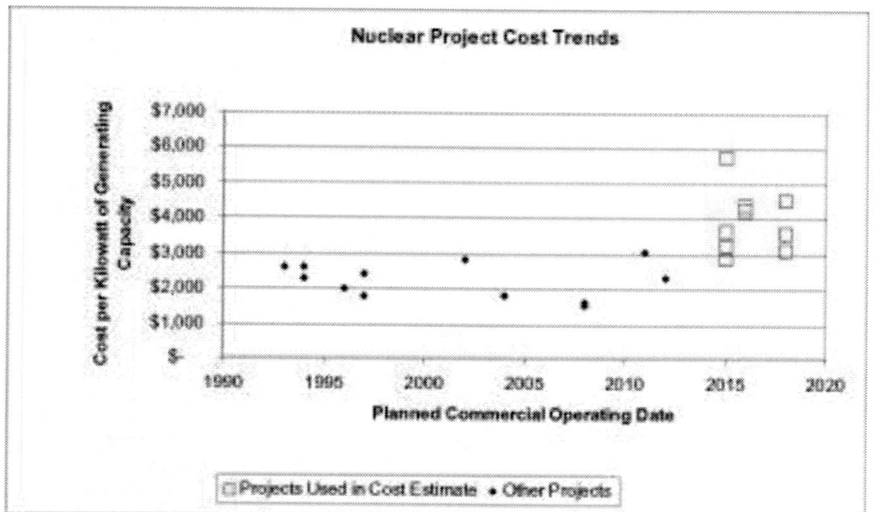

Natural Gas Combined Cycle

Combined Cycle Projects Selected for Cost Estimate
(Average Cost per Kw: $1,165; Rounded Average: $1,200)

Plant Name	State	Lead Developer	Type of Ownership	Energy Source	Technology	Net Sum-mer Capaci ty (Mw)	Cost (million $)	Cost per Kw	COD Year	Greenfield (G) or Brownfield (B)	Sources
Green-land Energy Center	FL	JEA	Utility	NG	Combined Cycle	553	$600	$1,085	2012	G	David Hunt, "JEA Plans New Natural Gas Plant," The Florida Times-Union, June 27, 2008; JEA, "Proposed Power Plant: Greenland Energy Center" [www.jea.com]; Air Permit Application to the Florida Department of Environmental Protection, No. 03 10072-015.
Avenal Power Project	CA	Macquarie Energy No rth American Trading Inc.	IPP	NG	Combined Cycle	483	$530	$1,097	2012	G	Application of Avenal Power Center, LLC, submitted to the California Energy Commis-sion Docket No. 08-AFC-1, 2/13/08.
Cane Island Combined Cycle	FL	Florida Mun-icipal Power Agency	Utility	NG	Combined Cycle	300	$350	$1,167	2011	B	Florida Municipal Power Agency Press Re- lease, January 9, 2008.

(Continued)

Plant Name	State	Lead Developer	Type of Ownership	Energy Source	Technology	Net Summer Capaci ty (Mw)	Cost (million $)	Cost per Kw	COD Year	Greenfield (G) or Brownfield (B)	Sources
Colusa Gen-erating Sta-tion	CA	Pacific Gas & Electric Co.	Utility	NG	Combined Cycle	527	$673	$1,277	2010	G	Pacific Gas & Electric Co., Opening Brief before the Calif-ornia Public Utilities Commi-ssion, Docket A.07-1 1-009.
Deer Creek	SD	Basin Electric Power Coopera-tive	Utility	NG	Combined Cycle	300	$330	$1,100	2012	G	Basin Electric Power Cooperative, "Deer Creek Station Joins Basin Electric's Fleet," Basin Today, November/December 2007.
Harry Allen Combined	NV	Nevada Power	Utility	NG	Combined Cycle	484	$682	$1,409	2011	B	Nevada Public Utilities Com-mission Docket No. 08-03-034: Application of Nevada
Cycle											Power; Direct Testimony on Behalf of Nevada Power of William Rodgers, Roberto Denis, and John Lescenski.
Thetford	MI	Consumers Energy	Utility	NG	Combined Cycle	512	$521	$1,017	2011	B	Direct testimonies of Lyle Thornton and Michael Torrey, on behalf of Consumers Energy Co., before the Michigan Public Service Commission, Case U-1 5290, May 1, 2007.

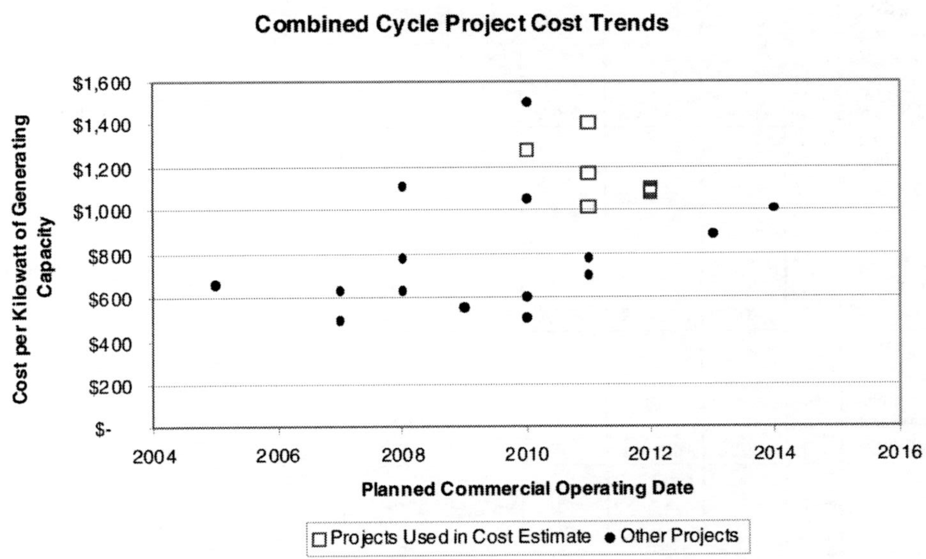
Combined Cycle Project Cost Trends
$1,600
$1,400
$1,200
$1,000
$800
$600
$400
$200
$-
Cost per Kilowatt of Generating Capacity
2004
2006
2008
2010
2012
2014
2016
Planned Commercial Operating Date
Projects Used in Cost Estimate
Other Projects

Wind

Wind Projects Selected for Cost Estimate (Average Cost per Kw: $2,106; Rounded Average: $2,100)

Plant Name	State	Lead Developer	Type of Ownership	Energy Source	Technology	Net Summer Capacity (Mw)	Cost (million $)	Cost per Kw	COD Year	Greenfield (G) or Brownfield (B)	Sources
Taconite I Wind Energy Center	MN	Minnesota Power	Utility	Renewable	Wind Turbine	25	$50	$2,000	2008	G	Minnesota Power Co., Petition for Approval, Minnesota Public Utilities Commis- sion Docket E01 5/M-07-1 064, August 3, 2007.
Blue Sky Green Field Wind Project	WI	Wisconsin Electric Power Co.	Utility	Renewable	Wind Turbine	145	$313	$2,152	2008	G	Final Decision, Wisconsin Public Service Commission, Application of Wisconsin Electric Power Co., Docket 6630-CE-294, February 1, 2007; WEPCO Second Quarter 2007 Progress Report, File 6630-CE-294, July 30, 2007.
Cedar Ridge Wind Farm	WI	Wisconsin Power and Light	Utility	Renewable	Wind Turbine	68	$165	$2,439	2008	G	Alliant Energy web site, cces-sed 2/5/2008 [http://www.alliantenergy.com/docs/groups/public/documents/pub/p01 5392.hcsp#P78_15008]; Alliant Energy press release, July 2, 2007; Alliant Second Quarter 2007 Progress Report, Docket

(Continued)

Plant Name	State	Lead Developer	Type of Ownership	Energy Source	Technology	Net Summer Capacity (Mw)	Cost (million $)	Cost per Kw	COD Year	Greenfield (G) or Brownfield (B)	Sources
											6680-CE-1 71, October 31, 2007; Wisconsin Public Service Commission, Certificate and Order, Docket 6680-CE-171, May 10, 2007.
Cloud County Wind Farm and Flat Ridge Wind Farm	KA	Westar Energy	Utility	Renewable	Wind Turbine	149	$269	$1,806	2008	G	Kansas State Corporation Commission, Final Order, Docket 08-WSEE-309-PRE, December 27, 2007; Direct Testimony of Greg A. Greenwood, Westar Energy, Docket 08-WSEE-309-PRE, October 1, 2007;
White Wind Farm	SD	Navitas Energy	IPP	Renewable	Wind Turbine	200	$300	$1,500	2010	G	Wayne Ortman, "South Dakota: State Utilities Commission Approves Permit for $300 Million Wind Farm," Associated Press, June 26, 2007; 2010 COD date per telecon with Doug Copeland of Navitas, 2/12/2008.

(Continued)

Plant Name	State	Lead Developer	Type of Ownership	Energy Source	Technology	Net Summer Capacity (Mw)	Cost (million $)	Cost per Kw	COD Year	Greenfield (G) or Brownfield (B)	Sources
Bent Tree Wind Farm	MN	Wisconsin Power and Light	Utility	Renewable	Wind Turbine	200	$463	$2,313	2010	G	Alliant Energy press release, June 6, 2008; Application of Wisconsin Power & Light before the Wisconsin Public Service Commission, Docket 6680-CE-173, June 6, 2008.
Crane Creek Wind Project	IA	Wisconsin Public Service	Utility	Renewable	Wind Turbine	99	$251	$2,535	2009	G	Wisconsin Public Service Commission, Certificate and Order, Docket 6690-CE194, May 22, 2008; Wisconsin Public Service Commission, letter amending Certificate and Order, Docket 6690-CE-194, May 28, 2008.

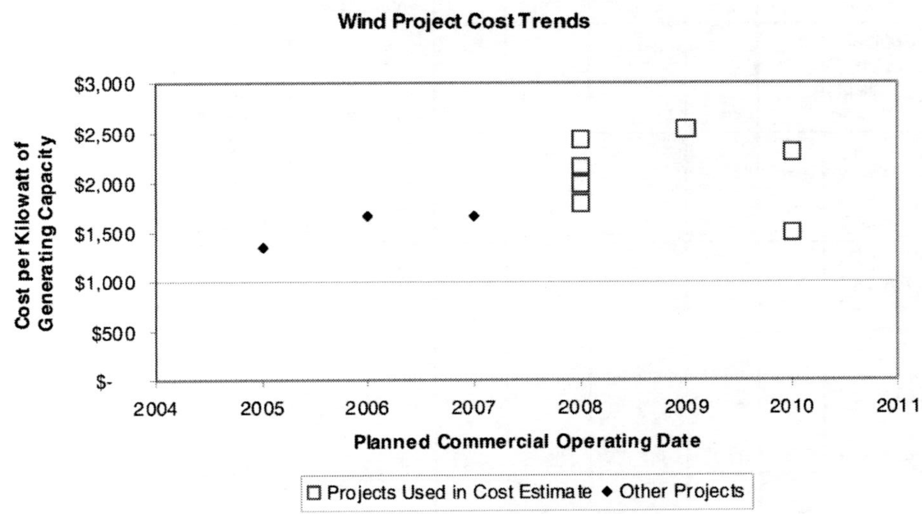
Wind Project Cost Trends
Cost per Kilowatt of Generating Capacity
$3,000
$2,500
$2,000
$1,500
$1,000
$500
$-
2004
2005
2006
2007
2008
2009
2010
2011
Planned Commercial Operating Date
Projects Used in Cost Estimate
Other Projects

Geothermal

Geothermal Projects Selected for Cost Estimate (Average Cost per Kw: $3,170; Rounded Average: $3,200)

Plant Name	State	Lead Developer	Type of Ownership	Energy Source	Technology	Net Summer Capaci ty (Mw)	Cost (million $)	Cost per Kw	COD Year	Greenfield (G) or Brownfield (B)	Sources
Newb-erry Volcano Pro ject (Phase I and II)	OR	Northwest Geo-thermal	IPP	Renewable	Geothermal	120	$300	$2,500	2011	G	Cindy Powers, "Suit Means Lik-ely Delays in Proposed Geo-thermal Plant," The (Bend,Or- egon) Bul-letin, 121/21/2006; Gail Kinsey Hill, "Company Set to Probe Cra-ter Area for Geothermal Project," The (Port-land, Oregon) Oregon-ian, 11/29/2007; [http://www.new berrygeothermal.com/ project.htm].
Faulkner I (Blue Moun-tain)	NV	Nevada Geo-thermal Power	IPP	Renewable	Geothermal	35	$120	$3,429	2009	G	"Nevada Geothermal Power Arra-nges $120 ml Financing to Begin 35-MW Project in Nevada," Plat-ts Global Power Report,8/2/2007.
Raft River Phase I	ID	U.S. Geo-thermal	IPP	Renewable	Geothermal	14	$39	$2,847	2008	B	Robert Peltier, "Renewable Top Plants," Power Magazine, Decem-ber 2007; EERE Network News, 1/9/2008.
Hot Sulfur Springs	NV	Fortis Capital	IPP	Renewable	Geothermal	32	$125	$3,906	2009	G	Thomas Rains, "EIF Dishes Out Lead Slots for Western Projects," Power, Finance and Risk, 12/14/2007.

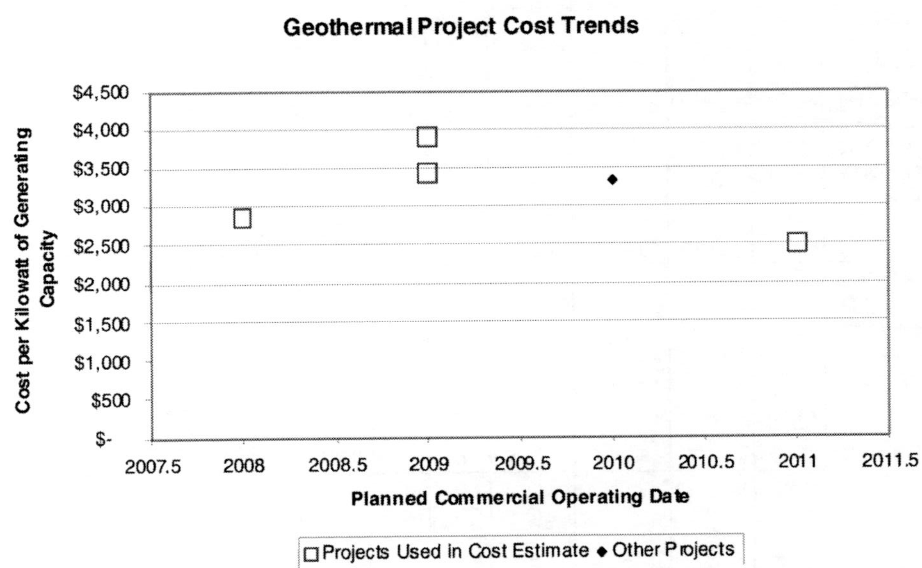
Geothermal Project Cost Trends
$4,500
$4,000
$3,500
$3,000
$2,500
$2,000
$1,500
$1,000
$500
$-
Cost per Kilowatt of Generating Capacity
2007.5
2008
2008.5
2009
2009.5
2010
2010.5
2011
2011.5
Planned Commercial Operating Date
Projects Used in Cost Estimate
Other Projects

Solar Thermal

Solar Thermal Projects Selected for Cost Estimate (Average Cost per Kw: $3,436; Rounded Average: $3,400)

Plant Name	State	Lead Developer	Type of Ownership	Energy Source	Technology	Net Summer Capacity (Mw)	Cost (million $)	Cost per Kw	COD Year	Greenfield (G) or Brownfield (B)	Sources
Bethel	CA	Bethel Energy 1 and 2	IPP	Renewable	Thermal PT	99	$368	$3,725	2008	G	Katy Burne, "California Solar Platform Nears Stake Sales," wer, Finance and Risk, October 5, 2007; "Project Finance Deal Book," Power, Finance and Risk, January 26, 2007; California Public Utilities Commission, Resolution E-4073, March 15, 2007.
Ivanpah	CA	BrightSource Energy	IPP	Renewable	Thermal Tower	400	$1,200	$3,000	2012	G	Peter Maloney, "Solar Power Heats Up, Fueled by Incentives and the Prospects of Utility-Scale Projects," Platts Global Power Report, November 1, 2007; "Storage: Solar Power's Next Frontier," Platts Global Power Report, November 1, 2007; California Energy Commission, Ivanpah Solar Electric Generating System Licensing Case, Docket 07-AFC-05 [http://www.energy.ca.gov/ sitingcases/ ivanpah/index.html].

(Continued)

Plant Name	State	Lead Developer	Type of Ownership	Energy Source	Technology	Net Summer Capacity (Mw)	Cost (million $)	Cost per Kw	COD Year	Greenfield (G) or Brownfield (B)	Sources
Carrizo En-ergy Solar Farm	CA	Ausra Inc.	IPP	Renewable	Thermal Other	177	$550	$3,107	2012	G	"PG&E Signs PPA for 177-MW Solar Project by Ausra in San Luis Obispo County, Calif.," Platts Global Power Report, November 8, 2007; California Energy Commission, Carrizo Energy Solar Farm Power Plant Licensing Case, Docket 07-AFC-08 [http://www.energy.ca].
Nevad a Solar	NV	Acciona	IPP	Renewable	Thermal	64	$250	$3,906	2007	G	Robert Peltier, "Renewable Top Plants,"
One		Solar Power			PT						Power Magazine, December 2007.
Mojave Solar Park	CA	Solel Solar Systems	IPP	Renewable	Thermal PT	554	$2,000	$3,610	2011	G	Terence Chea, "PG&E to Buy Electricity from Massive Solar Park in Mojave Desert," Associated Press, July 26, 2007; California Public Utilities Commission, Resolution E4138, December 20, 2007.
Xcel Solar Therm	CO	Xcel Energy	Uti-lity	Renewable	Thermal UNK	200	$600	$3,000	2016	G	Steve Raabe, "Big Solar Generator Proposed by Xcel," The Denver Post, November 16, 2007.

Table. (Continued)

Plant Name	State	Lead Developer	Type of Ownership	Energy Source	Technology	Net Summer Capacity (Mw)	Cost (million $)	Cost per Kw	COD Year	Greenfield (G) or Brownfield (B)	Sources
FPL Group Florida	FL	Florida Power & Light	Utility	Renewable	Thermal Other	300	$900	$3,000	2014	G	"FPL Plans to Build 300-MW Solar Project in Florida and Expand California Plant by 200 MW," Platts Global Power Report, September 27, 2007
Beacon Solar Energy Project	CA	Florida Power & Light Energy, LLC	IPP	Renewable	Thermal PT	250	$1,000	$4,000	2011	G	"FPL Plans to Build 300-MW Solar Project in Florida and Expand California Plant by 200 MW," Platts Global Power Report, September 27, 2007; California Energy Commission Fact Sheet, Beacon Solar Energy Project (08-AFC-2).
Solana Generating Station	AZ	Arizona Public Service	Utility	Renewable	Thermal PT	280	$1,000	$3,571	2011	G	Ryan Randazzo, "Plant to Brighten State's Solar Future," The Arizona Republic, 2/21/2008; http://www.aps.com/Solana; Thomas F. Armistead, "Arizona Utility Aims High for Solar Array," Engineering News-Record, 2/28/08.

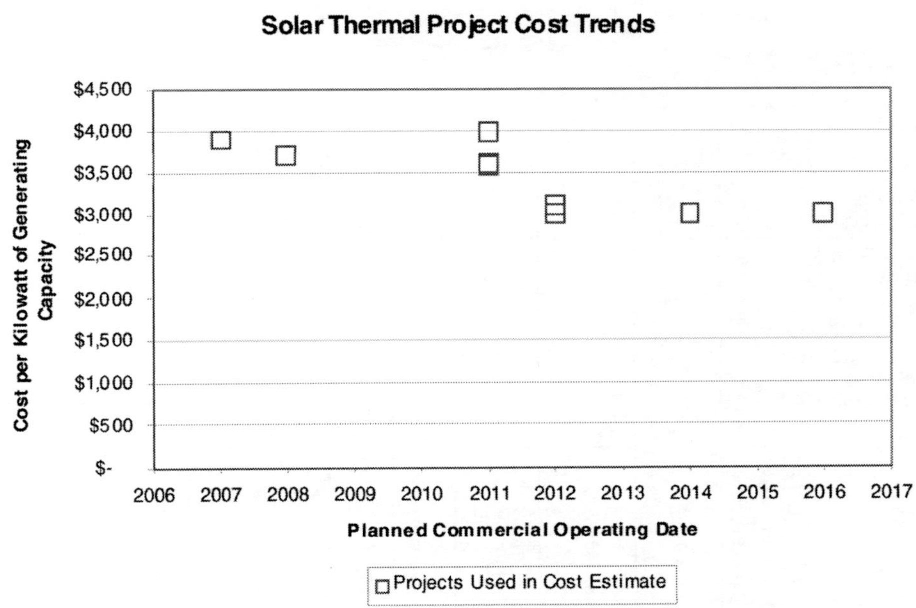
Solar Thermal Project Cost Trends
Cost per Kilowatt of Generating Capacity
$4,500
$4,000
$3,500
$3,000
$2,500
$2,000
$1,500
$1,000
$500
$-
2006
2007
2008
2009
2010
2011
2012
2013
2014
2015
2016
2017
Planned Commercial Operating Date
Projects Used in Cost Estimate

Solar Photovoltaic

Solar Photovoltaic (PV) Projects Selected for Cost Estimate (Average Cost per Kw: $6,552; Rounded Average: $6,600)

Plant Name	State	Lead Developer	Type of Ownership	Energy Source	Technology	Net Summer Capacity (Mw)	Cost (million $)	Cost per Kw	COD Year	Greenfield (G) or Brownfield (B)	Sources
Nellis Air Force Base	NV	MMA Renewable Ventures	IPP	Renewable	PV	14	$100	$7,143	2007	G	Tony Illia, "North America's Largest PV Powerplant in Service," Engineering News- Record, December 21, 2007; Nevada Power Press Release, December 17, 2007; John G. Edwards, "Photovoltaic Installation Finished at Air Force Base," Las Vegas Review Journal, December 18, 2007.
Alamosa Photovoltaic Power Plant	CO	SunEdison, LLC	IPP	Renewable	PV	8	$49	$5,961	2007	G	Erin Smith, "PUC Approves SunEdison Plant," Knight Ridder Tribune Business News, February 10, 2007.

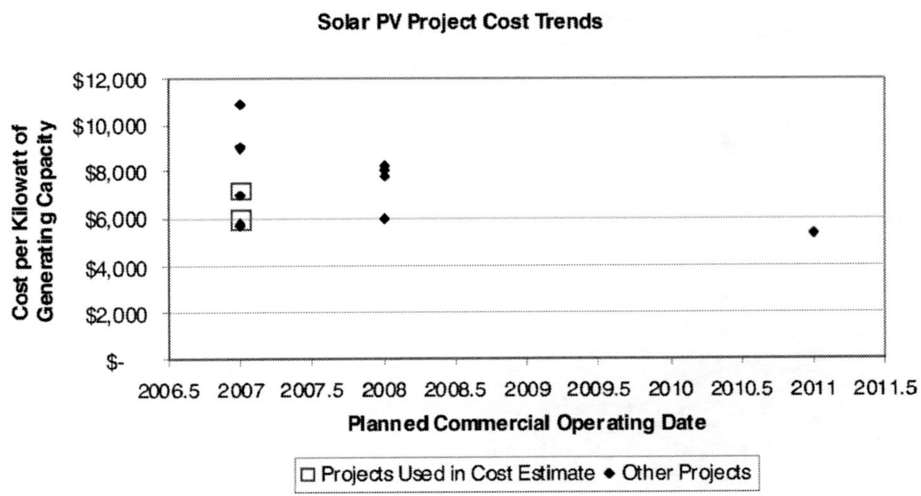
Solar PV Project Cost Trends
Cost per Kilowatt of Generating Capacity
$12,000
$10,000
$8,000
$6,000
$4,000
$2,000
$-
2006.5
2007
2007.5
2008
2008.5
2009
2009.5
2010
2010.5
2011
2011.5
Planned Commercial Operating Date
Projects Used in Cost Estimate
Other Projects

Appendix C

ESTIMATES OF TECHNOLOGY COSTS AND EFFICIENCY WITH CARBON CAPTURE

Pulverized Coal with Carbon Capture

The costs and heat rate for a supercritical pulverized coal plant with carbon capture is primarily based on information from MIT's 2007 study, *The Future of Coal.*[97] MIT estimated that a new supercritical plant built with amine scrubbing for CO_2 removal would have the following characteristics:

- CO_2 capture rate: 90%
- Change in efficiency compared to a new plant without carbon capture: -23.9% (from 38.5% to 29.3%). This equates to an increase in the heat rate of 31.3%.
- Increase in capital cost: 61%.[98]

For a new plant with amine scrubbing to have the same 600 MW net capacity as a new plant without carbon controls, the size of the plant has to be scaled up to account for the electricity and steam demands of the capture system. The increase is proportional to the change in efficiency. Therefore, a developer would have to build the equivalent of a 788 MW plant with carbon capture to get 600 MW of net capacity, with the difference (188 MW) consumed by the amine scrubbing system, either in the form of steam diverted from power generation or electricity used to compress the CO_2.[99]

MIT does not break out the variable and fixed O&M costs for carbon capture, as required by the financial model used in this study. These costs were calculated from a DOE study of the costs of retrofitting carbon capture to the Conesville Unit 5 coal-fired plant in Ohio. Based on this study, the incremental O&M costs for carbon capture are $8.24 per kW for fixed O&M and $7.79 per Mwh for variable O&M (2006 dollars).[100] These costs for operating the carbon capture system are added to the base O&M costs for a coal-fired plant, as estimated by EIA, to calculate the total O&M costs for the plant.

The estimated characteristics of a new supercritical pulverized coal plant with amine scrubbing are:

- Capacity: 600 MW.
- Heat rate: the base heat rate of 9,200 btus per kWh in 2008 increases by 3 1.3% to 12,080 btus per kWh.
- Overnight capital cost: $4,025 per kW (base 2008 cost of $2,500 per kW increased by 6 1%).
- Variable O&M costs (2006 dollars): a base value of $5.86 per Mwh plus the carbon control incremental cost of $7.79 per Mwh for a total of $13.65 per Mwh.
- Fixed O&M costs (2006 dollars): a base of $35.20 per kW plus the carbon control incremental cost of $8.24 per kW for a total of $43.44 per kW.[101]
- Capacity factor: 85%, same as for a new supercritical plant without carbon capture.
- Construction time: assumed to be four years, same as for a new supercritical plant without carbon capture.

IGCC Coal and Natural Gas Combined Cycle with Carbon Capture

The operating and cost characteristics of a coal IGCC plant built with carbon capture are taken from EIA assumptions for its 2008 long-term forecast,[102] except for the capital cost. As shown in Appendix B, the cost estimate for an IGCC plant without carbon capture, based on public information on current projects, is $3,400 per kW in 2008. This is much higher than EIA's estimate for an IGCC plant without ($1,773 per kW) or with ($2,537) carbon controls.

To estimate the capital cost of an IGCC plant with carbon capture, the percentage difference in the EIA estimates of plants with and without capture (43%) was applied to the CRS estimate of $3,400 per kW without capture. This produces an estimated cost for an IGCC plant with carbon controls of $4,862.[103] EIA's other assumptions, such as for O&M costs and heat rates, are used without adjustment in this study.

The capital cost for a natural gas-fired combined cycle with carbon capture was estimated in the same way. Based on public data for current projects, the overnight cost estimate for a new combined cycle used in this study is $1,200 per kW in 2008 (see Appendix B). This compares to EIA's estimates of $706 per kW for a combined cycle without carbon capture and $1,409 with carbon capture, a premium of 100%.[104] The capital cost for a new combined cycle with carbon capture used in this study is therefore double the CRS base cost of $1,200 per kW, or $2,400 per kW. As with the coal IGCC, EIA's other assumptions for a combined cycle plant with carbon capture are used without adjustment.

Appendix D

FINANCIAL AND OPERATING ASSUMPTIONS

Table 17. Financial Factors

Item	Value	Sources and Notes
Representative Bond Interest Rates		
Utility Aa	2010: 6.8% 2015: 7.0% 2020: 7.0%	When available, interest rates for investment grade bonds with a rating of Baa or higher (i.e., other than high yield bonds) are Global Insight forecasts. When Global Insight does not forecast an interest rate for an inv-estment grade bond the value is estim-ated based on historical relation-ships between bond interest rates (the his-torical data for this analysis is from the Global Finance website). High yield interest rates are estimated ba-sed on the differential between Mer-rill Lynch high yield bond indices and corporate Baa rates, as report-ed by WSJ.com (Wall Street Jour-nal website).
IPP High Yield	2010: 9.8% 2015: 10.0% 2020: 10.0%	
Public Power Aaa	2010: 5.1% 2015: 5.4% 2020: 5.4%	
Public Power Times Interest Earned Ratio Requirement	25%	
Corporate Aaa	2010: 6.3% 2015: 6.5% 2020: 6.5%	
Cost of Equity — Utility	14.00%	California Energy Commission, *Comparative Cost Of California Cental Station Electricity Generating Technologies*, December 2007, Table 8.
Cost of Equity — IPP	15.19%	
Debt Percent of Capital Structure	Utility: 50% IPP: 60% Utility or IPP with federal loan guarantee: 80% POU: 100%	Northwest Power and Conservation Council, *The Fifth Northwest Electric Power and Conservation Plan*, May 2005, Table I-1.

Table 17. Financial Factors

Item	Value	Sources and Notes
Federal Loan Guarantees		
Cost of equity premium for entities using 80% financing.	1.75 percentage points	Congressional Budget Office, *Nuclear Power's Role in Generating Electricity*, May 2008, web supplement ("The Methodology Behind the Levelized Cost Analysis"), Table A-5 and page 9.
Credit Subsidy Cost	12.5% of loan value	
Long-Term Inflation Rate (change in the implicit price deflator)	1.9%	Global Insight
Composite Federal/State Income Tax Rate	38%	EIA, National Energy Modeling System Documenta-tion, Electricity Market Module, March 2006, p. 85.

Notes: EIA = Energy Information Administration; IOU = investor owned utility; POU = publicly owned utility; IPP = independent power producer. For a summary of bond rating criteria see [http://www.bondsonline.com/Bond_Ratings_Definitions.php]. "High yield" refers to bonds with a rating below Baa.

Table 18. Power Plant Technology Assumptions (2008 $)

Energy Source	Technology	Overnight Construction Cost for Units Entering Service in 2015, 2008$ per kW[a]	Capacity (MW)	Heat Rate for Uni-ts Entering Service in 2015 (Btus per kWh)	Variable O& M Cost, 2008$ per Mwh	Fixed O&M, 2008$ per Megawatt	Capacity Factor
Pulverized Coal	Supercritical	$2,485	600	9,118	$4.68	$28,100	85%
Pulverized Coal: CC Retrofit	Subcritical	$2,192 (cost for CC retrofit only; original plant cost assumed to be paid off)	351	15,817	$16.15	$56,609	85%
Pulverized Coal:CC, New Build	Supercritical	$3,953	600	11,579	$14.32	$45,564	85%
IGCC Coal	Gasification	$3,359	550	8,528	$2.98	$39,459	85%
IGCC Coal: CC	Gasification	$4,774	380	10,334	$4.53	$46,434	85%
Nuclear	Generation II/III+	$3,682	1,350	10,400	$0.50	$69,279	90%
Natural Gas	Combined	$1,186	400	6,647	$2.05	$11,936	70%
Natural Gas: CC	Combined	$2,342	400	8,332	$3.00	$20,307	85%

Table 18. (Continued)

Energy Source	Technology	Overnight Construction Cost for Units Entering Service in 2015, 2008$ per kW[a]	Capacity (MW)	Heat Rate for Units Entering Service in 2015 (Btus per kWh)	Variable O& M Cost, 2008$ per Mwh	Fixed O&M, 2008$ per Megawatt	Capacity Factor
Wind	Onshore	$1,896	50	Not Applicable	$0.00	$30,921	34%
Geothermal	Binary	$3,590	50	Not Applicable	$0.00	$168,011	90%
Solar Thermal	Parabolic Trough	$2,836	100	Not Applicable	$0.00	$57,941	31%
Solar Photovoltaic	Solar Cell	$5,782	5	Not Applicable	$0.00	$11,926	21%

Sources: Heat rates, O&M costs, and nominal plant capacities are generally from the assumptions to EIA's 2008 Annual Energy Outlook; also see the other tables in this Appendix. Capital cost estimates are based on a CRS review of public information on current projects except for plants with carbon capture; see Appendix B. Capital costs and heat rates are adjusted based on the technology trend rates used by EIA in the Annual Energy Outlook, except for wind (cost is held constant between 2007 and 2010, instead of the increase EIA shows due to site specific factors). EIA costs are adjusted to 2008 dollars using Global Insight's forecast of the implicit price deflator. Capacity factor for coal plants is from MIT, *The Future of Coal*, 2007, p. 128. Natural gas plants without carbon capture are assumed to operate as baseload units with a capacity factor of 70%; natural gas with carbon capture operates at an 85% capacity factor, based on the assumption that such a plant would not be built other than to operate at a high utilization rate. Capacity factor for wind from California Energy Commission, *Comparative Costs of California Central Station Electricity Generation Technologies*, December 2007, Appendix B, p. 67. Nuclear plant capacity factor reflects the recent industry average performance as reported in EIA, *Monthly Energy Review*, Table 8.1. Capacity factors for solar and geothermal from EIA, Assumptions to the Annual Energy Outlook 2008, Table 73.

Notes: CC = carbon capture; kWh = kilowatt-hour; Mwh = megawatt-hour.

a. Construction costs include the affect of cost reductions due to technology improvements from the 2008 base levels reported in Appendix B.

Table 19. Air Emission Characteristics

Energy Source	Technology	Controlled SO_2 Emission Rate (pounds per MMBtu)	Controlled NOx Emission Rate (pounds per MMBtu)	CO_2 Emissions without Carbon Control (pounds CO_2 per MMBtu)	CO_2 Emissions with 90% Removal (pounds CO_2 per MMBtu)
Pulverized Coal	SupercriTical Pulverized Coal	0.157	0.05	209.0	20.9
IGCC Coal	Coal Gasification	0.0184	0.01	209.0	20.9
Natural Gas	Combined Cycle	0 (no controls required	0.02	117.08	11.708

Sources: DOE, Electric Power Annual 2006, Table A3; DOE, *20% Wind Energy by 2030,* May 2008, Table B-12; MIT, The Future of Coal, 2007, p. 139.

Notes: MMBtu = million btus; SO_2 = sulfur dioxide; NOx = nitrogen oxides; CO_2 = carbon dioxide. Coal emission rate for CO_2 is for a generic product computed as the average of the rates for bituminous and subbituminous coal.

Table 20. Fuel and Allowance Price Projections (Selected Years)

	Delivered Fuel Prices, Constant 2008$ per MillionBtus			Air Emission Allowance Price, 2008$ per Allowance		
	Coal	Natural Gas	Nuclear Fuel	Sulfur Dioxide	Nitrogen Oxides	Carbon Dioxide
2010	$1.93	$7.51	$0.73	$249	$2,636	*2012*: $17.70
2020	$1.80	$6.41	$0.78	$1,074	$3,252	$31.34
2030	$1.87	$7.48	$0.79	$479	$3,360	$63.99
2040	$1.96	$9.17	$0.76	$158	$3,180	$130.66
2050	$2.06	$11.24	$0.73	$52	$3,009	$266.80

Sources: Forecasts other than carbon dioxide allowances are from the assumptions to the Energy Information Administration's 2008 Annual Energy Outlook (AEO). Carbon dioxide allowance prices are from the backup spreadsheets for EIA's "Core" case analysis of S. 2191 [http://www.eia.doe.gov/ oiaf/servicerpt/s2 19

1/index.html]. The original values in 2006 dollars were converted to 2008 dollars using the Global Insight forecast of the change in the implicit price deflator. The EIA forecasts are to 2030; the forecasts are extended to 2050 using the 2025 to 2030 growth rates. The sulfur dioxide allowance forecast is for the western U.S., which is the best representation of national prices following the D.C. Circuit Court decision vacating the Clean Air Interstate Rule (which would have, in effect, created a premium for eastern region SO_2 allowances). The nitrogen oxides allowance forecast is for the eastern region of the United States, the only region for which an EIA forecast is available in the AEO output spreadsheet.

Notes: Btu = British thermal unit. Sulfur dioxide and nitrogen oxides allowances are dollars per ton of emissions; carbon dioxide allowances are dollars per metric ton of CO_2.

Appendix E

LIST OF ACRONYMS AND ABBREVIATIONS

ABWR	Advanced Boiler Water [nuclear] Reactor
AP1000	Advanced Passive 1000 [nuclear reactor]
BACT	Best Available Control Technology
CAIR	Clean Air Interstate Rule
CO	Carbon Monoxide
CO_2	Carbon Dioxide
CSP	Concentrated Solar Power
CWIP	Construction Work in Progress
DOE	U.S. Department of Energy
EIAEnergy	Information Administration
EOR	Enhanced Oil Recovery
EPRI	Electric Power Research Institute
ESBWR	Economic Simplified Boiling Water [nuclear] Reactor
Gen III/III+	Generation III/III+ (i.e., advanced) nuclear power plants
HAP	Hazardous Air Pollutant
IGCC	Integrated Gasification Combined Cycle
IOU	Investor Owned Utility;
IPP	Independent Power Producer
ITC	Investment Tax Credit
kW	Kilowatt
kWh	Kilowatt-hour
LAER	Lowest Achievable Emission Rate
LNG	Liquified Natural Gas

MACT	Maximum Available Control Technology
MIT	Massachusetts Institute of Technology
MMBtu	Millions of British Thermal Units
MW	Megawatt
Mwh	Megawatt-hour
NA	Not Applicable
NAAQS	National Ambient Air Quality Standards
NEI	Nuclear Energy Institute
NETL	National Energy Technology Laboratory
NM	Not Meaningful
NO_x	Nitrogen Oxides
O&M	Operations and Maintenance
POU	Publicly Owned Utility
PT	Parabolic Trough
PTC	Production Tax Credit
PV	Photovoltaic
RTO	Regional Transmission Organization
SCPC	Supercritical Pulverized Coal
SCR	Selective Catalytic Reduction
SO_2	Sulfur Dioxide
UNK	Unknown
U.S. – EPR	United States - Evolutionary Pressurized [nuclear] Reactor
USCPC	Ultra-Supercritical Pulverized Coal

End Notes

[1] EIA, an independent arm of the Department of Energy, is the primary public source of energy statistics and forecasts for the United States. The estimated amount of new generating capacity is taken from the Excel output spreadsheet for the *Annual Energy Outlook 2008* report. Note that EIA forecasts assume no change to the laws and regulations in effect at the time the forecasts are made.

[2] Variable costs are costs that vary directly with changes in output. For fossil fuel units the most important variable cost is fuel. Solar and wind plants have minimal or no variable costs, and nuclear plants have low variable costs.

[3] A combustion turbine is an adaption of jet engine technology to electric power generation. A combustion turbine can either be used stand-alone as a peaking unit, or as part of a more complex combined cycle plant used to meet intermediate and baseload demand.

[4] This alignment of generating technologies is for new construction using current technology. The existing mix of generating units in the United States contains many exceptions to this alignment of load to types of generating plants, due to changes in technology and economics. For instance, there are natural gas and oil-fired units built decades ago as

baseload stations that now operate as cycling or peaking plants because high fuel prices and poor efficiency has made them economically marginal Some of these older plants were built close to load centers and are now used as reliability must-run (RMR) generators that under certain circumstances must be operated, regardless of cost, to maintain the stability of the transmission grid.

[5] Hydroelectric generation is a special case. Hydro generation is very low cost and is firm, dispatchable capacity to the degree there is water in the dam's reservoir. However, operators have to consider not only how much water is currently available, but how much may be available in upcoming months, and competing demands for the water, such as drinking water supply, irrigation, and recreation. These factors make hydro dispatch decisions very complex. In general hydro is used to meet load during high demand hours, when it can displace expensive peaking and cycling units, but if hydro is abundant it can also displace baseload coal plants.

[6] For example, a solar project developer decided to leave storage and other "extras" out of a proposed plant in order to make it "commercially viable." "Storage: Solar Power's Next Frontier," *Platts Global Power Report,* November 1, 2007.

[7] There are different measures of capacity. Nameplate capacity is the nominal maximum output of a generator, and gross capacity is the actual maximum output. Net capacity is gross output minus the electricity needed to operate the plant. Net capacity is therefore the amount of capacity that can actually put electric power on the grid. Net capacity can vary with air and water temperatures, so a further distinction is made between summer and winter net capacity. Capacity factor is most commonly computed using net summer capacity.

[8] The estimate of 86% of 2006 generation from large baseload and intermediate generating units was computed from the EIA-860 (generating capacity) and EIA-906/920 (generation) data files for 2006, available at [http://www.eia.doe.gov/cneaf/electricity/page/data.html]. The calculation assumed that plants with a capacity factor of 25% or greater fall into the intermediate/baseload category, and that plants with a capacity of 200 MW or greater are "large." These thresholds are assumptions because there are no official categorizations of what constitutes intermediate, baseload, or large power plants. However, large changes to the threshold values do not change the conclusion. For example, if the capacity factor floor for what constitutes intermediate/baseload generation is increased to 33%, the intermediate/baseload percentage of generation is 83%; if the size threshold is increased to 300 MW, the intermediate/baseload percentage of generation is also 83%; and if both changes are made the intermediate/baseload percentage of generation is 81%.

[9] Generation from petroleum products dropped from 365.1 billion kilowatt-hours (kWh) in 1978 to 65.7 billion kWh in 2007. Almost a quarter of the 2007 petroleum generation came not from liquid fuels, such as distillate fuel oil, but from a solid refinery waste product, petroleum coke. EIA, *Annual Energy* Review 2006, Table 8.2a, and *Electric Power Monthly,* March 2008, Table ES1 .B.

[10] In 2007 total generation was 4,160 million Mwh. Generation from the industrial and commercial sectors totaled 154 million Mwh, some of which was from non-CHP industrial and commercial generators. EIA, *Annual Energy Review* 2007, Table 8.1.

[11] North American Electric Reliability Corp., *2008 Long-Term Reliability Assessment,* October 2008, p. 46.

[12] The primary alternative to pulverized coal technology for new coal plants is the circulating fluidized bed (CFB) boiler. CFB is a commercial system used mainly for relatively small scale plants (about 250 MW and less) that burn waste products (such as petroleum coke, a refinery residue) as well as coal. CFB is currently a niche technology and is not covered further in this report. For additional information see Steve Blankinship, "CFB: Technology of the Future?," *Power Engineering,* February 2008. (The article is available online by searching at [http://pepei.pennnet.com/]).

[13] EIA estimates a heat rate advantage of 4.7% for current technology. With projected improvements the difference widens substantially, to almost 15%. EIA, *Assumptions to the Annual Energy Outlook 2008*, Table 38. Another study is less optimistic, finding that IGCC "electricity generating efficiencies demonstrated to date do not live up to earlier projections due to the many engineering design compromises that have been made to achieve acceptable operability and cost. The current IGCC units have and next-generation IGCC units are expected to have electricity generating efficiencies that are less than [i.e., worse than] or comparable to those of supercritical P[ulverized] C[oal] generating units." Massachusetts Institute of Technology (MIT), *The Future of Coal,* 2007, p. 124.

[14] For instance, LS Power, a coal project developer, describes IGCC technology as "experimental." Steve Raabe, "'Clean Coal' Plant Setbacks Mount in U.S.," *The Denver Post,* November 1, 2007.

[15] For example, Appalachian Power (APCo, a subsidiary of the large utility American Electric Power) has proposed building an IGCC plant to serve customers in Virginia and West Virginia. The Virginia State Corporation Commission rejected the proposal, citing the technical immaturity and uncertain costs of IGCC technology. The same project was approved by the West Virginia Public Service Commission, which concluded that "the Project is an efficient and capable proposal to meet the baseload needs of APCo's customers" and is the "best option" available to APCo. (Virginia State Corporation Commission, Application of Appalachian Power Co., Case No. PUE-2007-0068, Final Order, April 14, 2008, pp. 12-13; West Virginia Public Service Commission, Application for a Certificate of Public Convenience and Necessity, Case No. 06-0033-E-CN, Commission Order, March 6, 2008, p. 25.)

[16] According to the 2006 version of the EIA-860 data file of generating units, between 1995 and 2006, inclusive, 255,980 MW of new generating capacity of all types entered service. Out of this total, 168,800 MW used generating technologies suitable for baseload and intermediate service, including geothermal, combined cycle, fuel cell, hydroelectric, steam turbines using combustible fossil or renewable fuels, and wind turbines. Of this baseload/intermediate segment, 148,119 MW was gas-fired combined cycles, or 88%. The next largest shares were wind power (6%) and coal (4%).

[17] EIA, *Annual Energy Outlook 2008*, p. 68; Matthew Wald, "Utilities Turn From Coal to Gas, Raising Risk of Price Increases," *The New York Times*, February 5, 2008; "FERC's Moeler Just Wants to Make it Clear: Natural Gas 'Fuel of Choice' in the Near Future," *Platts Electric Utility Week*, October 22, 2007; Alexander Duncan, "Power Needs, Climate Concerns to Spark 'Bullish' Natural Gas Market: Experts," *Platts Inside Energy*, October 8, 2007

[18] Calculated from the *Annual Energy Outlook 2008* output spreadsheet. EIA projects that natural gas-fired combined cycle plants plus natural gas combustion turbine peaking plants will account for 54% of capacity additions through 2015.

[19] Ibid. EIA projects the construction of 85,300 MW of new coal fired capacity.

[20] Rebecca Smith, "Banks Hope to Expand Carbon Rules to Public Utilities," *The Wall Street Journal*, March 20, 2008.

[21] DOE/NETL, *Tracking New Coal-Fired Power Plants*, June 2008, p. 5. This report is periodically updated and posted at [http://www.netl.doe.gov/coal/refshelf/ncp.pdf].

[22] North American Electric Reliability Corp., *2008 Long-Term Reliability Assessment*, October 2008, p. 88.

[23] According to the EIA-906/920 data file for 2007, gas-fired combined cycles accounted for 688 million megawatt-hours of generation, out of a total of 4,160 million megawatt-hours.

[24] For an illustrated summary of several of the Gen III/III+ designs, see "UK Nuclear Power: The Contenders," *BBC News,* January 10, 2008 [http://news.bbc.co.uk/2/hi/science/nature/ 51651 82.stm]. Additional information is available from the links at [http://www.nei.org/ keyissues/newnuclearplants/newreactordesigns/].

[25] As of August 2008, a reported 95 geothermal projects with publicly known generating capacities were in development in the United States. The upper estimate of the total capacity of these projects was 3,959.7 MW, or an average of 42 MW per project. All the projects are located in western states except for a single 1 MW project in Florida. Kara Slack, *U.S. Geothermal Power Production and Development Update,* Geothermal Energy Association, August 2008, p. 8.

[26] For additional information on geothermal power see Steve Blankinship, "What Lies Beneath," *Power Engineering,* January 2007, available by searching [http://pepei.pennnet.com/]).

[27] EIA, *Annual Energy Outlook 2008,* p. 70. For more detail on wind power, see CRS Report RL34546, *Wind Power in the United States: Technology, Economic, and Policy Issues,* by Jeff Logan and Stan Kaplan.

[28] For a comprehensive list of energy market incentives, see EIA, *Federal Financial Interventions and Subsidies in Energy Markets 2007*, April 2008.

[29] The analysis does not include the credit for carbon dioxide sequestration established by P.L. 110-343, Division B, Title I, Subtitle B, Section 115 (adding a new §45Q to 26 U.S.C.). The law provides for tax credits of $20 per metric ton of CO_2 sequestered and $10 per metric ton for CO_2 captured and used for enhanced oil recovery. The credit is in effect through the year in which the cumulative volume of CO_2 captured totals 75 million metric tons. This credit is excluded because it is very difficult to predict how long the credit will be in effect. The EIA analysis of the Lieberman-Warner Climate Security Act of 2009 (S. 2191) estimates, for the cases that project carbon capture, cumulative CO_2 capture of about 80 million to 100 million tons by 2014, which is prior to the on-line data of 2015 assumed for new power plants in this study. (For the spreadsheets which contain the detailed S. 2191 outputs, see the EIA website at [http://www.eia.doe.gov/oiaf/servicerpt/s2 19 1/index.html].)

[30] 26 U.S.C. §45, as amended by P.L. 110-343, Division B, Title I, Subtitle A, Section 101(a).

[31] 26 U.S.C. §45J.

[32] For a discussion of the operation of the credit see EIA, *Annual Energy Outlook* 2007, p. 21. For the forecast of 8,000 MW of nuclear capacity on-line before 2021, see the *Annual Energy Outlook 2008,* p. 70.

[33] 10 CFR § 609 (RIN 1901-AB21), October 4, 2007 [http://www.lgprogram.energy.gov/keydocs.html].

[34] On the assumption that the guaranteed debt would have a high (AAA) rating, see "Loan Guarantees for Projects that Employ Innovative Technologies," 10 CFR § 609 (RIN 1901-AB21), October 4, 2007, p. 24.

[35] Entities receiving loan guarantees must make a substantial equity contribution to the project's financing. Public power entities normally do not have the retained earnings needed to make such payments. The rules also preclude granting a loan guarantee if the federal guarantee would cause what would otherwise be tax exempt debt to become subject to income taxes. Under current law this situation would arise if the federal government were to guarantee public power debt. For further information on these and other aspects of the loan guarantee program see U.S. DOE, final rule, "Loan Guarantees for Projects that Employ Innovative Technologies," 10 CFR § 609 (RIN 1901-AB21), October 4, 2007 [http://www.lgprogram.energy.gov/keydocs.html].

[36] *DOE Announces Plans for Future Loan Guarantee Solicitations,* Department of Energy press release, April 11, 2008. According to press reports, the Japanese and French governments may also offer loan guarantees to American nuclear projects. French and Japanese companies are expected to be major suppliers to new U.S. nuclear projects. The terms of the loan guarantees, assuming they come to fruition, are unknown. Elaine Hiruo, "Japanese Government Considers Loan Guarantees for U.S. Reactors," *Platts Nucleonics Week,* August 14, 2008, and Elaine Hiruo, "Japan Clears Way for Loan Guarantees in US," *Platts Nucleonics Week,* September 25, 2008

[37] Steven Dolley, "Nuclear Power Key to Exelon's Low-Carbon Plan," *Platts Nucleonics Week* (February 14, 2008). For similar comments see "House Appropriators Seek DOE Loan

Guarantees Delay Pending GAO Review," EnergyWashington.com, June 10, 2008; Dr. Joe C. Turnage, UniStar Nuclear, presentation to the California Energy Commission, "New Nuclear Development: Part of the Path Toward a Lower Carbon Energy Future," June 28, 2007; and Selina Williams, "US Government Loan Guarantees For New Nuclear Too Small NRC," CNNMoney.com, March 10, 2008.

[38] 26 U.S.C. §48, as amended by P.L. 110-343, Division B, Title I, Subtitle A, Section 103(a)(1).

[39] For additional information see the discussion of the investment tax credit in the federal incentives section of the Database of State Incentives for Renewable Energy website [http://www.dsireusa.org/].

[40] Investor owned utilities did not qualify for this credit until the passage of P.L. 110-343 in October 2008. See P.L. 110-343, Division B, Title I, Subtitle A, Sections 103(e) and 103(f)(4).

[41] 26 U.S.C. §48A, as amended by P.L. 110-343, Division B, Title I, Subtitle B, Section 111.

[42] The IGCC credit is 20% capped at $133.5 million per project, with a requirement that the credits be allocated to projects in each of three categories: Bituminous coal-fired, subbituminous coal-fired, and lignite-fired plants. Other advanced coal technologies can qualify for a 15% credit (with a cap of $125 million per project) if 1) a new unit can achieve a heat rate of 8,530 btus/kWh or less and near zero non-CO2 emissions, or 2) an existing plant can meet various criteria for improving thermal efficiency, including by replacing inefficient old units at a plant site with new units.

[43] "Consumers Energy Latest to Win Tax Concessions," *Platts Electric Power Daily*, November 29, 2007.

[44] Mary Powers, "Governor Expected to Sign Mississippi Bill on Collecting Costs of Building Baseload," *Platts Electric Utility Week*, April 21, 2008; Elaine Hiruo and Tom Harrison, "Summer Owners Lock in Price, Schedule for Planned New Reactors," *Platts Nucleonics Week,* May 29, 2008. In addition, Florida, Louisiana, Virginia, and North Carolina will reportedly allow return on CWIP for nuclear plants (Dr. Joe C. Turnage, UniStar Nuclear, "New Nuclear Development: Part of the Strategy for a Lower Carbon Energy Future," presentation to the Center for Strategic and International Studies meeting "Evaluating the Business Case for Nuclear Power," July 31, 2008, p. 4). The treatment of CWIP in rates varies by jurisdiction and by case. The amount of CWIP allowed is typically updated periodically and may be limited by a total project cost approved by the commission

[45] Wisconsin Public Service Commission, Certificate and Order, Docket 6680-CE-1 71, May 10, 2007 (for Wisconsin Power & Light's Cedar Ridge project, estimated to cost $179 million); Kansas State Corporation Commission, Final Order, Docket 08-WSEE-309-PRE, December 27, 2007 (for Westar Energy's investment in the Central Plains and Flat Ridge wind projects, estimated to cost the utility $282 million).

[46] Typical practice is for the project developer to enter into a single EPC contract with a large construction and engineering firm. The firm is responsible for most plant construction activities and absorbs significant cost, delay, and technical risk, which is reflected in the contract price. A developer can act as its own EPC manager and avoid paying the risk premium to a third party contractor, but in this case the developer absorbs the price and performance risks.

[47] IHS CERA press release, "Construction Costs for New Power Plants Continue to Escalate IHS-CERA Power Capital Costs Index," May 27, 2008 [http://energy.ihs.com/News/ Press-Releases/2008/IHS-CERA-Power-Capital-Costs-Index.htm].

[48] Keith Bradsher and David Barboza, "Pollution From Chinese Coal Casts a Global Shadow," *The New York Times*, June 11, 2006.

[49] Christopher D. Kirkpatrick, "A Bidding War for Engineers: Power Plant Construction Boom Creates a Labor Shortage," *The Charlotte (North Carolina) Observer*, September 5, 2008.

[50] Yuliya Chernova, "Change in the Air," *The Wall Street Journal,* February 11, 2008; Bert Caldwell, "BPA's wind power tops 1,000 megawatts," *The (Spokane, Washington) Spokesman-Review,* January 12, 2008; Yoshifumi Takemoto and Alan Katz, "Samurai-

Sword Maker's Reactor Monopoly May Cool Nuclear Revival," Bloomberg.com, March 13, 2008.

[51] Matthew L. Wald, "Costs Surge For Building Power Plants," *The New York Times,* July 10, 2007.

[52] Wind power is less costly to build than, for example, coal or nuclear plants. However, because wind plants are weather dependent, wind plants have much lower capacity factors than coal or nuclear plants. A typical wind plant capacity factor is about 34%, compared to 70% to over 90% for coal and nuclear plants. This means the capital costs of a wind plant are spread over relatively few megawatt-hours of generation, increasing the cost per unit of electricity sold. In the case of variable renewable resources like wind and solar power, anything that reduces capital costs or increases utilization can significantly improve plant economics.

[53] For example, vendors in Asia and Europe are planning to add new capacity to manufacture very large forgings, particularly important for nuclear plants. Mark Hibbs, "Chinese Equipment Fabricators Set Ambitious Capacity Targets," *Platts Nucleonics Week,* May 22, 2008; Pearl Marshall, "UK's Sheffield Forgemasters Plans to Produce Ultra-large Forgings," Platts Nucleonics Week, April 3, 2008.

[54] Equity capital includes the funds provided by the owners of the firm (i.e., the stockholders). Debt is borrowed money. The owners of a project seek to repay debt, and to both recover their equity investment and earn a return on that investment.

[55] Prior to the restructuring of the electric power industry that began in the 1990s, IOUs were typically vertically integrated, providing generation, transmission, and distribution (final delivery of electricity to consumers) in a state-sanctioned monopoly service area. With restructuring, some states required or encouraged utilities to divest their power plants. In many parts of the country control (though not ownership) of transmission assets is now in the hands of federally sponsored regional transmission organizations (RTOs). Some states that required IOUs to divest generation are now allowing utilities to once again own and operate power plants, such as California.

[56] In 2006, out of 2,010 government-owned electric utilities, only 98 had total revenues in excess of $100 million dollars. In contrast, the fuel cost for a single large power plant can exceed $100 million per year. American Public Power Association, *2008-09 Annual Directory and Statistical Report,* p. 30 (data does not include electric cooperatives).

[57] In some parts of the country RTOs operate power markets and have capped spot electricity prices, such as at $1,000 per Mwh, to prevent extraordinary price spikes. These caps apply to spot sales of electricity, not to bilateral contracts.

[58] Because the debt is tax free, the POU can pay the bond holder a lower interest rate than taxable debt must offer. The bond holder accepts the lower POU tax-free interest rate since, other things being equal, its after-tax return is the same.

[59] Moody's Investors Service, *Mapping of Moody's U.S. Municipal Bond Rating Scale to Moody's Corporate Rating Scale and Assignment of Corporate Equivalent Ratings to Municipal Obligations*, June 2006, p.2. According to Moody's, between 1970 and 2000, out of 699 rated municipal bond issues for electric power, only two defaulted (including the Washington Public Power Supply System default on a large nuclear construction program). Over the same period, about 70% of municipal bonds were rated A or higher, and less than 1% were rated below investment grade. Moody's Investors Service, *Moody's US Municipal Bond Rating Scale*, November 2002, pp. 5-6.

[60] Moody's Investors Service, *Moody's US Municipal Bond Rating Scale*, November 2002, p. 6. Rating agencies assign debt to credit worthiness categories. Investment grade debt has a rating of BBB- or higher in the nomenclature used by Standard & Poors and Fitch. The equivalent category for Moody's is Baa3 and higher. Lower rated debt is referred to as speculative or high yield issues, or less pleasantly as "junk bonds." For descriptions of the ratings systems and crosswalks see Edison Electric Institute, *2007 Financial Review*, p. 86, and [http://www.nnnsales.com/faq/faq-buyersinvestors8.htm]. Note that the municipal bond

market was roiled by the 2008 financial crisis (Tom Herman, "Muni Yields Rise to Rare Levels" *The Wall Street Journal*, November 5, 2008).

[61] Roughly 70% of utility companies were rated between BBB+ and BBB- in 2007. About 10% were rated below investment grade. Edison Electric Institute, *2007 Financial Review*, pp. 81 and 87.

[62] Most IPP debt is reportedly rated below investment grade (telephone conversation with Scott Solomon, Moody's Investors Service, February 15, 2008). For instance, in June 2008 the debt ratings for several large IPP developers were all speculative grade: NRG (Standard & Poors B rating), AES (B+ to BB-), Edison Mission Energy (BB-), and Dynegy (B-). (Source: Standard & Poors NetAdvantage on-line data system). IPP power plants may be project-financed; that is, the financing and the recourse of the debt holders is tied to a specific project, not to the corporation as a whole. For example, the LS Power Sandy Creek, AES Ironwood, and Calpine's Riverside and Rocky Mountain projects all have project-specific, speculative grade debt ratings. (Source: Moody's Investors Service press releases, August 3, 2006, August 14, 2007, and February 8, 2008.)

[63] Over-reliance on debt is considered risky for private entities and leads investors to demand higher interest rates. At some level of debt a project would be impossible to finance. POUs can rely on 100% debt financing because they control their own rates and are backed-up by the government entity that owns or finances the utility.

[64] Equity is more expensive than debt in part because interest payments on debt are tax deductible while the imputed cost of equity is not an expense for income tax purposes. Another consideration is that in the event of bankruptcy bondholders are paid before shareholders. An equity investment is therefore riskier than holding debt and investors demand higher compensation. (Unlike a bond which has a known interest rate, there is no directly measurable cost of equity. Its cost is essentially the return investors will expect on their equity stake in the firm. Various techniques are used to estimate the cost of equity. The concepts are discussed in standard finance texts; see for example, Stewart Myers and Richard Brealey, *Principles of Corporate Finance,* 7th edition, 2003, Chapter 9.)

[65] Financing arrangements can be far more complex than described in this brief overview. As an illustration, see the discussions of wind power financing in Ryan Wiser and Mark Bolinger, *Annual Report on U.S. Wind Power Installation, Cost, and Performance Trends: 2007*, U.S. DOE, May 2008, p. 14; and John P. Harper, Matthew D. Karcher, and Mark Bolinger, *Wind Project Financing Structures: A Review & Comparative Analysis,* Lawrence Berkeley Laboratory, September 2007. For a description of the financing arrangements for an IPP-developed coal plant, see the discussion of the Plum Point project in "North American Single Asset Power Deal of the Year 2006," *Project Finance,* February 2007.

[66] Coal and gas prices have increased due to national and global demand growth, limited excess production capacity, certain unusual circumstances (such as flooding that reduced Australian coal production and exports), increases in rail, barge, and ocean-going vessel rates for delivering coal to consumers, and the run-up in world oil prices. For a discussion of energy price trends, see EIA's *Annual Energy Outlook* for long-term projections and the *Short-Term Energy Outlook* for near-term forecasts [http://www.eia.doe.gov/oiaf/forecasting .html].

[67] EIA, *Annual Energy Outlook Retrospective Review*, April 2007, p. 5.

[68] Ibid., pp. 2 and 3 [table citations omitted].

[69] Factors that caused prices to rise include increased demand, problems bringing new uranium mines into service, and the depletion of commercial inventories of uranium. The recent decline in prices may be due in part to an improved short-term production outlook; see "ERI Expects Base Price to Drop, Then Rise Again," *Platts Nuclear Fuel*, June 16, 2008. It takes years before a change in uranium prices is reflected in a reactor fuel load. The lag is caused by the time it takes to process the uranium and manufacture fuel rods; multi-year contracts that do not reflect current prices; and reactor fueling schedules (refueling takes place on 18 or 24 month cycles, and at each refueling only about a third of the core is replaced). This lag

can cut both ways: If uranium prices decline, a plant may still have reloads based on expensive uranium in the pipeline.

70 For the EIA nuclear fuel price forecast used in the *Annual Energy Outlook 2008*, go to [http://www.eia.doe.gov/oiaf/aeo/electricity.html] and click on "figure data" for Figure 70.

71 Under the existing federal SO_2 and NOx regulatory programs, most existing plants have been allocated allowances sufficient to cover their emissions. These existing plants do not need to buy emissions, and may have surplus emissions to sell, especially if the plants have retrofitted pollution control equipment.

72 Coal plants can produce two types of particulates. Primary particulates, sometimes referred to as soot, are formed in the combustion process. Secondary particulates form in the atmosphere through the condensation of nitrates and sulfates. Particulates are objectionable because of visibility and health effects. For more information see Rod Truce, Robert Crynack, and Ross Blair, "The Problem of Fine Particles," *Coal Power*, September 30, 2008 [http://www.coalpowermag.com/environmental/156.html].

73 Renewable power plants that do not burn fuels, such as solar, wind, and geothermal power, do not have air emissions. The depleted fuel rods from nuclear plants contain high level radioactive wastes. The nuclear fuel costs used in this study include the federal one mill (i.e., one tenth of a cent) per kWh fee for supporting creation of a permanent waste repository. In the interim depleted fuel is stored at each reactor site. For more information see CRS Report RL3 3461, *Civilian Nuclear Waste Disposal,* by Mark Holt.

74 BACT requirements take into account cost-effectiveness; LAER requires the lowest possible emission rate without cost considerations. For an overview of the regulatory framework see MIT, *The Future of Coal,* 2007, pp. 135 - 136. The federal New Source Performance Standards for new, large fossil-fired plants are found at 40 C.F.R. §60(Da).

75 An allowance is authorization to emit one unit of a pollutant during a specified time period, usually a year. For example, under the acid rain cap and trade program, national total SO_2 emissions are capped and each coal plant must submit sufficient allowances to cover its annual emissions. Older plants can comply by staying within emission allocations, installing control equipment, and/or buying SO_2 allowances. New plants must install control equipment *and* buy allowances.

76 NO_x regulation is complex and involves both federal and state rules. For a summary of NO_x regulation see the National Energy Technology Laboratory website at [http://www.netl.doe.gov/technologies/coalpower/ewr/nox/regs.html].

77 The decision has been appealed by the EPA to the U.S. Supreme Court.

78 RS22817, *The D.C. Circuit Rejects EPA's Mercury Rules: New Jersey v. EPA,* by Robert Meltz and James E. McCarthy; Amena Saiyid, "Utilities with Permits to Build New Units Caught in MACT Regulatory Bind," *Platts Coal Outlook*, June 23, 2008.

79 A 600 MW coal plant with an 85% capacity factor and a heat rate of 9,000 btus per kWh, will consume about 40.2 trillion btus of fuel per year. At a controlled emission rate of 0.157 lbs of SO_2 per million btus of fuel consumed, this results in emissions of about 3,200 tons of SO_2 annually. At a late June 2008, SO_2 allowance price of $330 per ton, this equals an annual cost of $1.1 million. Emissions and the resulting allowance cost would be still less for an IGCC. In contrast, the fuel cost for this hypothetical plant (assuming a delivered cost of Central Appalachian coal of $137.92 per ton and a heat content of 12,500 btus per pound) would be about $222 million per year. The SO_2 system does consume a material amount of the electricity produced by a pulverized coal plant, in the range of 1% to 3% of output. Sources: MIT, *The Future of Coal,* 2007, p. 138; Spark Spreads table, *Platts Coal Trader,* June 30, 2008; U.S. DOE, 20% *Wind Energy by 2030,* Table B-12; Delivered Coal Price Comparison table, *Argus Coal Transportation,* June 24, 2008.

80 There are also many CRS reports on climate change issues. These reports can be retrieved by using the "Energy, Environment, and Resources" link on the CRS home page to access the "Climate Change" link.

[81] Currently four commercial facilities in the United States treat fossil plant flue gas to recover CO_2. The largest amount of CO_2 captured is about 800 tons per day. In contrast, a 600 MW coal plant would produce about 13,300 tons of CO_2 daily; 90% removal would require extracting 12,000 tons of CO_2 each day. (Information on current commercial projects from HDR|Cummins & Barnard, Inc., *Carbon Dioxide Capture and Sequestration*, report to Alliant Energy, April 2008, Report No. 5561.06 R-002, p. 8; and [http://www.mgs.md.gov/geo/pub/co2seqpaper.pdf]. CO_2 emissions for a 600 MW plant computed as follows: 600 MW x 9 million btus of fuel input per MWh x 24 hours x 205.3 pounds of CO_2 released per mmbtu of heat input for bituminous coal, divided by 2 million. Rate of CO_2 released from burning coal is from EIA, *Electric Power Annual 2006,* p. 92.)

[82] MIT, *The Future of Coal,* 2007, pp. 25 and 28; "Pilot Project Uses Innovative Process to Capture CO2 From Flue Gas,' *EPRI Journal*, Spring 2008, p. 4).

[83] Calculated from MIT, *The Future of Coal,* 2007, Table 3.1 (estimates for supercritical pulverized coal).

[84] Ibid., p. 28. The cost and practicality of a retrofit would vary with specific plant conditions. Another consideration is that retrofitting carbon capture to an IGCC plant may not be straightforward. An MIT study suggests that for technical reasons a developer looking toward possible future carbon legislation cannot build an IGCC plant that will provide optimal efficiency today (without carbon technology) and tomorrow (after carbon control retrofit). The developer must make a choice that may result in suboptimal performance (higher costs and less efficiency) either in current or future operation (MIT, *The Future of Coal,* 2007, pp. 149-150).

[85] National Energy Technology Laboratory, *Cost and Performance Baseline for Fossil Energy Plants, Volume 1,* May 2007, Exhibit 5-25 and page 481; EIA, *Assumptions to the Annual Energy Outlook 2008,* Table 38. The plant capacity derate for the natural gas combined cycle plant is less than for the pulverized coal plant primarily because natural gas generation is much less carbon intensive than burning coal, so less CO_2 must be processed. The lower carbon intensity is due to the greater efficiency of a gas-fired combined cycle compared to a pulverized coal plant (fewer btus of fuel are needed to generate a unit of electricity), and because burning a btu of gas produces about half as much CO_2 as burning a btu of coal.

[86] The dry feed Shell and ConocoPhillips E-Gas systems appear to be better suited to high moisture subbituminous and lignite coals than the GE technology, which brings coal into the gasifier as a coal/water slurry (excess water reduces the efficiency of the gasifier and requires more oxygen). However, the GE technology operates at higher pressures and can use full quench cooling of the synthesis gas to produce steam for the CO_2 shift reactor, which may make it the better choice for carbon capture. MIT, *The Future of Coal*, 2007, pp. 149 - 151; EPRI, *Feasibility Study for an Integrated Gasification Combined Cycle Facility at a Texas Site*, October 2006, pp. v and vi; and Nexant, Inc., *Environmental Footprints and Costs of Coal-Based Integrated Gasification Combined Cycle and Pulverized Coal Technologies*, report for the U.S. EPA, July 2006, p. 5-13.

[87] For a broader summary of S. 2191 allowance price forecasts see CRS Report RL34489, *Climate Change: Costs and Benefits of S. 2191/S. 3036,* by Larry Parker and Brent D. Yacobucci. For an example of how a different legislative approach can effect allowance prices, see CRS Report RL34520, *Climate Change: Comparison and Analysis of S. 1766 and S. 2191 (S. 3036)*, by Larry Parker and Brent D. Yacobucci.

[88] For a more detailed discussion of the annualization method see, for example, Chan Park, *Fundamentals of Engineering Economics,* 2004, Chapter 6; or Eugene Grant, et al., *Principles of Engineering Economy,* 6th Ed., 1976, Chapter 7.

[89] For additional information on capital charge rates see Hoff Stauffer, "Beware Capital Charge Rates," *The Electricity Journal,* April 2006. For additional information on the calculation of capital recovery factors see Chan Park, *Fundamentals of Engineering Economics,* 2004, Chapter 2; or Eugene Grant, et al., *Principles of Engineering Economy,* 6th Ed., 1976, Chapter 4.

[90] The *Annual Outlook* main report, assumptions report, and related information are available on the EIA website at [http://www.eia.doe.gov/oiaf/aeo/index.html].

[91] George Lobsenz, "Nuke Overload: Utilities Seeking $122 Billion in DOE Loan Guarantees," *The Energy Daily*, October 3, 2008.

[92] Recent coal projects with public power participation include Prairie State (Illinois), Spruce 2 (Texas), Spurlock 4 (Kentucky), Dallman 4 (Illinois), Smith CFB (Kentucky), Sutherland 4 (Iowa), Pee Dee (South Carolina), Cross 3 and 4 (South Carolina), Whelan 2 (Nebraska), Hugo 2 (Oklahoma), Southwest 2 (Missouri), Dry Fork (Wyoming), Nebraska City 2 (Nebraska), Weston 4 (Wisconsin), Big Stone II (South Dakota), Plum Point (Arkansas), Turk (Arkansas), American Municipal Power Generating Station (Ohio), and Holcomb 2&3 (Kansas). Proposed new nuclear projects with POU involvement include Summer 2 and 3 (South Carolina), Vogtle 3 and 4 (Georgia), North Anna 3 (Virginia), Bellefonte 3 and 4 (Alabama), Calvert Cliffs 3 (Maryland), and South Texas 3 and 4 (Texas). Some of the coal projects and all of the nuclear projects other than Bellefonte have IOU or IPP co-owners. The POU participant in the Calvert Cliffs 3 project is EDF, a French government-owned utility.

[93] EIA, *Annual Energy Outlook 2008*, p. 75.

[94] Rebecca Smith, "Utilities Question Natural-Gas Forecasting — Cheap and Plentiful Was Outlook a Few Years Ago; Price Is Double Prediction," *The Wall Street Journal*, December 27, 2004.

[95] EIA, *Energy Market and Economic Impacts of S. 2191, the Lieberman-Warner Climate Security Act of 2007*, April 2008. The report and output spreadsheets are available at the EIA website at [http://www.eia.doe.gov/oiaf/servicerpt/s2 19 1/index.html]. Note that the carbon case in this report does not include other aspects of S. 2191 that would affect compliance costs, including a free allowance allocation and carbon control bonus allocations of allowances.

[96] The pulverized coal plant modeled in this study emits about 1,906 pounds of CO_2 per Mwh. This is computed as follows. The plant has a heat rate of 9,118 btus per kWh. This equates to coal consumption of 9.118 MMbtus per Mwh. Coal is assumed to emit 209 pounds of CO_2 per mmbtu of coal consumed, so 9.118 MMbtus per Mwh x 209 pounds of CO_2 per mmbtu = 1,905.7 pounds of CO_2 per Mwh. In the case of a combined cycle burning natural gas, the gas emits only 117.08 pound of CO_2 per mmbtu when burned (44% less than coal) and the plant's heat rate is 6,647 btus per kWh (27% better than the coal plant). The combined cycle's CO_2 emissions are therefore 6.647 MMbtus per Mwh x 117.08 pounds of CO_2 per mmbtu = 778.2 pounds of CO_2 per Mwh, 59.2% less than the pulverized coal plant.

[97] MIT, *The Future of Coal, 2007*, p. 30, Table 3.5.

[98] Another recent study shows a capital cost premium of 82%. DOE/National Energy Technology Laboratory, *Cost and Performance Baseline for Fossil Energy Plants, Volume 1*, May 2007, Exhibit 4-46.

[99] The required capacity is computed as 600 MW x (base efficiency of 3 8.5% / efficiency with carbon capture of 29.3%) = 788.4 MW.

[100] The DOE study estimates the incremental O&M costs for the carbon capture system. These costs, in 2006 dollars, are fixed O&M of $2.5 million per year and variable O&M of $17.6 million. The capacity of the unit after the installation of carbon capture is 303,317 kW, and the estimated capacity factor is 85%. The fixed O&M per kW is therefore $17.6 million / 303,317 kW = $8.24 per kW. The variable O&M per Mwh is $17.6 million / (303,317 x 85% x 8760 hours / 1000) = $7.79 per Mwh. DOE /National Energy Technology Laboratory, *Carbon Dioxide Capture from Existing Coal-Fired Power Plants*, DOE/NETL-401/110907, revised November 2007, pp. ES-3, 120, and 124.

[101] The base O&M values are derived from EIA, *Assumptions to the Annual Energy Outlook 2008*, Table 38. The EIA values must be adjusted because, as discussed above, the unit is in effect a 788 MW plant derated to 600 MW. The adjustment is proportional to the difference in efficiency between the plant with and without carbon capture, respectively 38.5% and

29.3%. The ratio of these values (1.314) is the adjustment factor. The adjusted fixed O&M cost is the EIA value of $26.79 per kW x 1.314 = $35.20. The adjusted variable O&M is the EIA estimate of $4.46 per Mwh x 1.3 14 = $5.86 per Mwh.

102 EIA, *Assumptions to the Annual Energy Outlook 2008*, Table 38.

103 MIT's cost estimates show a smaller capital cost premium of 32% for IGCC with and without carbon capture. MIT, *The Future of Coal, 2007,* p. 30, Table 3.5. A DOE study shows a premium range of 32% to 40%, depending on the type of IGCC system assumed. DOE/National Energy Technology Laboratory, *Cost and Performance Baseline for Fossil Energy Plants, Volume 1, 2007,* Exhibit 3-114.

104 The EIA data is from *Assumptions to the Annual Energy Outlook 2008,* Table 38. A DOE study estimates a cost premium of 112%. DOE/National Energy Technology Laboratory, *Cost and Performance Baseline for Fossil Energy Plants, Volume 1, 2007*, Exhibit 5-25.

INDEX

A

B

C

D

E

F

G

H

I

J

K

L

M

N

O

P

R

S

T

U

V

W

Y